THE
CLOCKMAKERS
OF LONDON

An account of the Worshipful Company of Clockmakers,
its Library and its Collection.

George White

An early example of a pocket watch by John Arnold (1736–1799) containing his pivoted detent escapement. Signed on its movement 'John Arnold invt. et Fecit No. 28'. It was completed c.1777.
Museum No. 419

FOREWORD

When the author was asked to reorder the Clockmakers' Museum in the City of London in the late 1990s, he was determined not to duplicate the display of any other horological collection. The Victoria and Albert Museum, he felt, was there to demonstrate design, the Royal Observatory to explain navigation, and the Science Museum to explain technology. What approach therefore should the Clockmakers' Museum take? The answer he decided, was 'people'. The Company, founded in 1631, consists (and has always consisted) of people. It was they who made the majority of objects in its collection and they whose story the museum should tell.

The first edition of this book was designed to accompany the exhibition that resulted. This second, fully revised and re-photographed edition is published to celebrate the move of the exhibition to a much larger space in the Science Museum in South Kensington.

'Many carry Watches about them that do little heed the fabric and contrivance, or the wit and skill of workmanship; as there be many that dwell in this habitable world that do little consider or regard the wheelwork of this great Machine, and the fabric of the house they dwell in.'

Humane Industry, London 1661

A superb silver-cased watch signed by James Vautrollier who worked 'without Temple Bar', London c.1625.
Museum No. 708

THE IMPORTANCE OF TIMEKEEPING AND OF LONDON AS A CLOCK- AND WATCHMAKING CENTRE

It is often thought that precision timekeeping is a relatively recent concept. This is based on the belief that past generations led a slower and less regimented existence, working to the nearest hour, with little use for minutes and seconds. In fact, the need for accurate timekeeping first became acute as early as the 16th century, when Europe's maritime nations began to look beyond their own shores, towards the riches of the New World and the Far East.

Navigators had long been able to discover their latitude while out of sight of land, but had never found a means of establishing their longitude with accuracy. It was known that one way of solving this frequently fatal problem was to carry precise and reliable timekeepers on all voyages, but no such instruments existed. It was from this early period that the search for technical solutions began and horology as a subject was propelled towards the forefront of scientific experiment and debate.

Similarly, it is often thought that clock- and watchmaking in past centuries was based exclusively in central Europe and that the finest work has always come from Switzerland. Again, this is not so. Many of the most significant inventions were either made or brought to perfection in England, in particular in London. It was the spectacular advances in English horology in the second half of the 18th century, consolidated in the 19th century, which enabled the British to explore the globe, map what they found, trade repeatedly with far-away places, conquer foreign lands and, in due course, acquire an empire.

THE BEGINNING OF CLOCKMAKING

Right:
The frontispiece of Sebastian Munster's *Horologio Graphia* (1533), an early horological treatise on the design of sundials and one of the treasures of the Clockmakers' Company Library.
Clockmakers' Library

The origin of the first mechanical clocks in Europe is as much disputed as the date at which they were made. Public clocks began to appear in the great cities and cathedrals from the end of the 13th century.

The first domestic clocks seem to have been made in Italy, France and Germany during the middle of the 15th century. The first watches, which developed through the miniaturisation of spring-driven clockwork, were made in the middle years of the 15th century.

The elements in mechanical clocks and watches, whether great or small, which give them their ability to keep time, are their 'escapements' and their 'controllers'. These are the parts which together regulate the speed at which the wheels turn when a

Right:
A small anonymous spring clock or watch, with a balance and verge escapement. Probably made in Germany in the second quarter of the 16th century.
Museum No. 582

weight or spring is applied to them. The real challenge to the scientific community in its search for greater accuracy was always to devise increasingly precise escapements and controllers, while at the same time reducing inconsistencies in the power train that kept them in motion.

From the earliest days until around 1660, no significant improvement in design of controller was achieved. Clocks and watches with balance or foliot controlled verge escapements kept adequate time for domestic purposes. Weight clocks, because of the uniformity of their power source, could be expected to keep better time than spring clocks or watches, but neither even remotely approached the accuracy required for calculating longitude at sea.

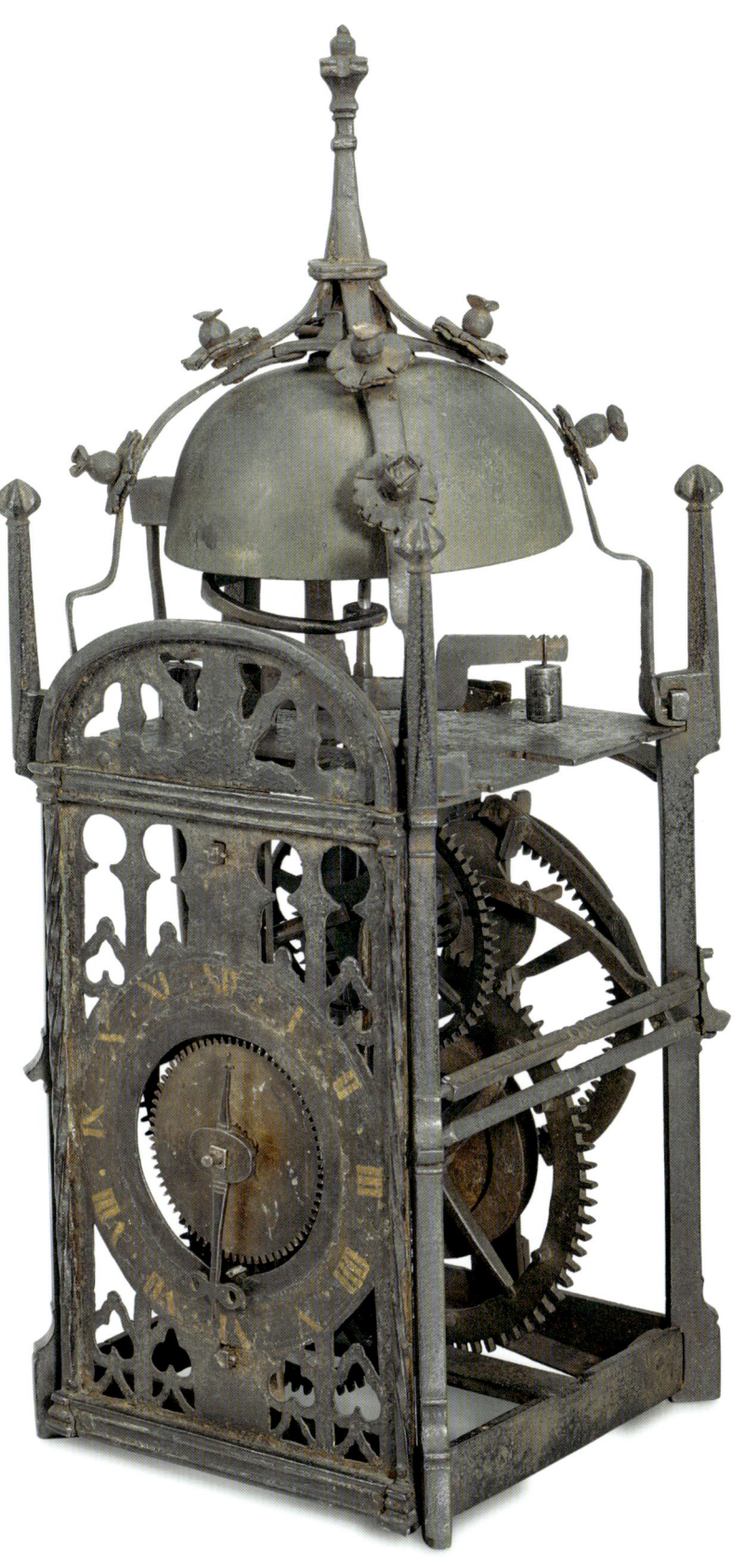

Right:
An anonymous 15th-century domestic clock with foliot and verge escapement, probably made in Germany.
Museum No. 573

THE BEGINNING OF CLOCKMAKING | 7

THE GUILD SYSTEM

Failure to improve timekeeping did not mean failure to increase horological skills. By the middle of the 16th century, central European clock and watchmakers began to produce work of an ingenuity and complexity unmatched before. Nowhere was this more so than in the cities of southern Germany, where two strong influences coincided. The first was the immense patronage from the Imperial and princely courts. Clocks and watches were ordered not only for scientific and domestic use and as 'toys' to ornament royal collections, but to embellish the huge annual tribute which had been payable by the Christian west from 1454 onwards to the Turks, to keep the Sultan and hoards at bay.

The second was the influence of the craft guilds, whose growing power in the southern German states allowed them to control quality, output and design with such vigour that only the best and most ingenious work ever reached the market place.

Quite how these guilds formed is uncertain, but it seems reasonable to suppose that they developed throughout Europe from the early medieval church guilds, or 'friendly societies', whose primary purpose was devotional, but whose secondary purpose was the welfare of souls, mutual society, and the care of the sick and of the poor. Initially they bore the names of saints and centred around the chapels of city and even country churches. It can only be assumed that men of common skills gravitated to the same chapels, especially within towns and so the trade-orientated urban guild system evolved.

Guilds dedicated specifically to clock- and watchwork, which had often broken away from earlier general metalworking or smiths' guilds, established themselves in the mid-16th century in cities such as Munich, Vienna, Innsbruck, Strasbourg, Nuremburg and Augsburg. Their rules and byelaws varied a little from city to city,

but in practice they had much the same effect. In Augsburg, for example, it was not possible to practice the art of clockmaking, without first serving an apprenticeship to a Master Craftsman. After three years' apprenticeship, a young man could become a journeyman. He remained a journeyman until he became eligible to apply to make his own masterpiece, which could be up to thirteen years later. Even then, he had to make it to rigid specifications which tested every aspect of his ability.

Left:
A superb automaton masterpiece made by Johann Schneider in Augsburg c.1625 to 16th-century guild rules. The specifications for horological masterpieces set in Augsburg in 1577 remained in force well into the 18th century, making the city's products increasingly antiquated and unsaleable.
Museum No. 585

THE GUILD SYSTEM | 9

Right:
Nicholas Kratzer, a Bavarian mathematician and horologist, encouraged to come to London by King Henry VIII in around 1516. In this portrait, after an original by Holbein, Kratzer is shown surrounded by the tools of his trade.
Private Collection

EARLY DOMESTIC HOROLOGY IN ENGLAND

Elsewhere in continental Europe, straightforward domestic clocks began to be produced in some numbers for the benefit of the rich, drawing on the inspiration of Italy and the southern German states. In cities such as Paris, Blois and Lyon, centres of watchmaking began to develop. In England, despite the encouragement that King Henry VIII, King Edward VI and Queen Elizabeth gave to foreign watch and clockmakers to settle in London, no distinctive tradition of native domestic horology immediately established itself. English smiths were well capable of producing public clocks. According to documentary sources, they occasionally turned their hands to the manufacture of smaller items too, but almost all the clocks and watches recorded in the inventories of major households in England from the early 16th until the following century, were imported or were the work of immigrant craftsmen.

Right:
The oldest London-made clock in the Clockmakers' Company Collection bears the simple signature 'Henry Archer' on its movement (right). It was made as late as c.1625. The case (left) is likely to have been imported from France. Archer was known to have been working in London by 1622. He was appointed as one of the Company's first Wardens in 1631 and served as its Deputy Master in 1632.
Museum No. 584
Photograph by the author

EARLY CLOCKMAKING IN LONDON

Above:
A fine early 17th-century watch by Robert Grinkin of Fleet Street. The movement includes an alarm, which rings on an oval bell within the case. Grinkin was a freeman of the Blacksmiths' Company and probably served as Master in 1608. He died in 1626.
Museum No. 6
Photograph by the author

The domination of the English clock and watch trade by immigrant craftsmen at the end of the 16th century was reinforced as religious persecution by the French and Spanish drove more and more Huguenot craftsmen from France and the Low Countries to safety across the channel. Many were attracted by the potentially rich pickings offered by the City of London, where they were obliged to buy themselves into one of the many existing craft guilds or City companies in order to gain access to the market. Some joined the Blacksmiths' Company, others joined any company that was prepared to take them. Yet others set up illegally either within the city or around its margins (known as 'the liberties'), where guild rules were hard to enforce and 'fines' or the subscriptions payable could be evaded.

Probably the best known among those immigrants today are Francis and Michael Nouwen (or Nowé), who are believed to have come from Brabant, and Nicholas Vallin, also from the Low Countries. These three, together with Bartholomew Newsam, perhaps a native Englishman, have come to be regarded as the fathers of the English domestic clock trade, because their work constitutes the earliest surviving examples that can clearly be shown to have been made in England. Early surviving watches by Michael Nouwen, Francis Nouwen and Randolph Bull perhaps put them in a similar position in terms of the watch trade.

Progress was cut short, however, by the tragedy of bubonic plague, which ravaged parts of the City in 1598 and 1603. Francis Nouwen died in the first attack. Newsam also died in the same year, whether by plague or not is not known. Vallin, his two daughters Margaret and Jane, and his two journeymen, John Archer and John Leyns, all died in the second attack, which was particularly virulent in and around St. Anne's parish in Blackfriars, where immigrants from the Low Countries had gathered. Six hundred and seventy of their community died in that year, thirty-three thousand Londoners died in all.

The consequence of this catastrophe seems to have been a period of retrenchment, especially in clockmaking, and a readjustment in the balance of nationality within the trade. English makers, with backgrounds in locksmithing, blacksmithing, needlemaking and similar metalworking skills, joined the continental makers who had survived, and distinctive London styles began to emerge. At the same time, the structure of the trade strengthened and developed, based on an

Left:
A very beautiful watch by the immigrant Michael Nouwen, who worked in London in the late 16th century. The blue enamel chapter ring encloses a dial centre perfectly engraved with hares and foliage. (The single hour hand is missing).
Museum No. 5

Left, background:
Many immigrant clock and watchmakers lived and worked in Blackfriars, including Nicholas Vallin and Francis Nouwen. Their community was devastated by the plagues of 1598 and 1603.
From the late 16th century 'Agas' map, reproduced by kind permission of Guildhall Library.

Actual size

interdependent group of specialist workers such as founders, engravers, manufacturers, retailers and wholesalers.

To some extent there was subdivision. Watches, and their larger relation, the spring clock, became the province of one group, whose members included Robert Grinkin, James Vautrollier, David Bouquet, David Ramsay, William Petit, Henry Archer, Edmund Bull, Simon Bartram and many others. They were heavily influenced by the French among them, who maintained close trading contacts with their families abroad.

Weight-driven clocks became the province of a second group of makers, some of whom were native English; others were immigrant stock, mostly originating from the Low Countries. They included Robert Harvey of Little Britain Street, who had come to London from Oxford around 1602, Henry Stevens, probably of Addling Hill, who was repairing London church clocks as early as 1599, Peter Closon of Holborne Bridge, William Bowyer of Leadenhall Street and Francis Foreman of St Paul's. They produced what they knew as 'house clocks' (now known as 'lantern clocks'), the staple domestic timekeepers of the period. Customers were the nobility, gentry and rich merchants. Among them were the earls of Arundel, Derby, Rothes and Marischal.

THE FOUNDING OF THE CLOCKMAKERS' COMPANY

Facing page top:
The minutes of the first meeting of the Court held on 12th October 1632. Present were David Ramsay, Henry Archer, John Willow, Sampson Shelton, James Vautrollier, John Smith, Francis Foreman, Samuel Linaker, John Chalton, John Midnall, Simon Bartram and Edward East.
MS 2710 Vol. 1

Facing page:
A house clock signed 'Francis Foreman, at St. Paules gate'. Foreman and Richard Morgan delivered the Clockmakers' petition to King Charles I in 1630.
Museum No. 1165

The success of the London trade brought problems because it attracted more and more craftsmen to the City, to the considerable annoyance of those already enjoying the benefits of working there. The London makers felt threatened by the continuing influx of outsiders and refugees and were especially annoyed by those who set up illegally or in the 'liberties', avoiding the expense and guild responsibilities that they themselves had to bear.

Twice, in 1620 and 1622, a group who claimed (somewhat mischievously) to be the only lawful clockmakers in the City, petitioned the Crown for permission to break away from the assortment of ancient guilds to which they were attached and form a specialist company of their own. To the great distress of those guilds and to the irritation of the Blacksmiths' Company in particular, they finally achieved their aim. In reply to a petition delivered to King Charles I in the hands of Francis Foreman and Richard Morgan, the King granted a charter for the formation of a Clockmakers' Company on 22nd of August 1631. The byelaws, or rules of the Company, followed in 1632, the combined documents giving the Company authority over the arts of clock and watchmaking, case making, sundial making, mathematical instrument making and engraving. Authority in most circumstances could only be exercised in the area covered by the City of London and ten miles beyond, but in relation to the inspection and hallmarking of imported goods, the area was extended to the whole of England and Wales.

THE FOUNDING OF THE CLOCKMAKERS' COMPANY | 15

Above:
The charter of incorporation granted to the Clockmakers' on 22nd August 1631 by King Charles I. The Company paid £4 to Mr John Chappell in 1634 for its 'flourishing and finishing'.
MS 6430

The structure of the new Company exactly mirrored those already in existence. Fourteen 'Assistants' were chosen from the membership or 'Freedom' to form the governing body or 'Court'. Three were chosen to serve as 'Wardens', whose duty was to serve the 'Master' and care for the Company's assets. Funds were raised by 'fines' or down payments and 'quarterage', which was a regular fee charged quarterly. The posts of Master and Wardens were subject to annual election. The system remains in operation to this day.

Actual size

Left and above (detail):
A watch by Henry Archer who was named as Warden by the Charter, but served as Deputy Master during the Company's first year.
Museum No. 20

THE FOUNDING OF THE CLOCKMAKERS' COMPANY | 17

The famous 'star watch' by the first Master of the Clockmakers' Company. Signed 'David Ramsay Scottes me fecit' (meaning 'David Ramsay the Scot made me') it was made c.1625. The silver case was exquisitely engraved by Gérard de Heck of Blois, whose name is recorded above twelve o'clock on the dial. The principal scenes show the birth of Christ, the Visitation, the Annunciation, the Nativity, the Adoration, and the Presentation in the Temple. The watch was discovered c.1794 at Gawdy Hall in Norfolk, where it had been concealed in a recess behind a tapestry, with two silver apostle spoons and documents dating from the time of the Civil War.

Museum No. 7

Actual size

18 | THE CLOCKMAKERS OF LONDON

David Ramsay (a Scot who had worked in France and followed James I to London) was appointed the first Master. As a courtier and a polymath with many other interests, he rarely attended, and Henry Archer, one of the Wardens, acted in his stead. Edward East, James Vautrollier, Francis Foreman and Simon Bartram were amongst those appointed Assistants in the early years.

The international nature of the guild system makes it hard to decide whether the byelaws of the new Company were based entirely on those of its fellow London companies, or whether specific notice was taken of the clockmaking guilds overseas. Either way there is no doubt that in horological particulars, the new Company was very similar to the earliest guilds in the southern German states.

In return for rigorous protection by their Company from outside competition, London clock- and watchmakers had to accept strict regulations of their own behaviour towards their customers,

THE FOUNDING OF THE CLOCKMAKERS' COMPANY | 19

Left:
An example of the work of 'our well-beloved James Vautrollier', one of the Company's first Assistants named in the Charter. The watch is magnificently engraved with a scene perhaps representing John 3:1-21, the meeting of Christ and Nicodemus. An angel above plays the English bagpipes.

Museum No. 708

Actual size

20 | THE CLOCKMAKERS OF LONDON

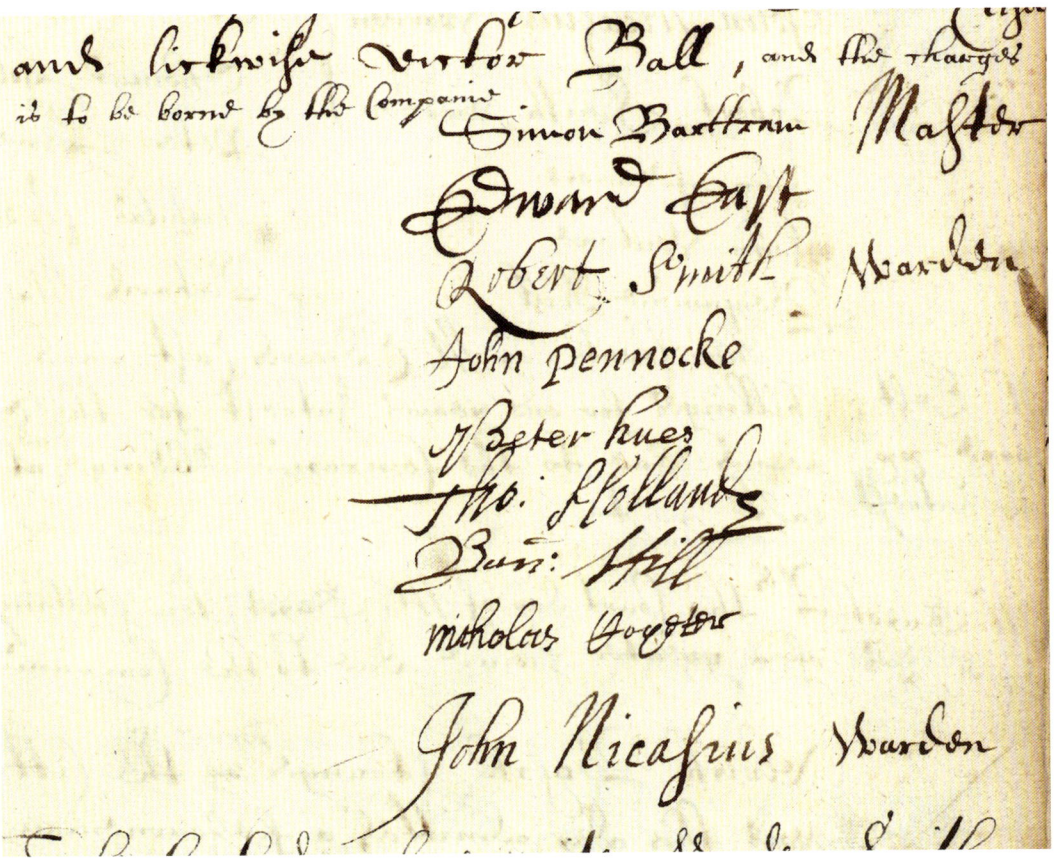

Left:
Signatures of some of the great mid-17th-century watch- and clockmakers, including Simon Bartram, Edward East, John Pennock, Benjamin Hill, Nicholas Coxeter and John Nicasius. Dated 1652.
MS 2711 Vol. 1

towards their fellow makers and towards their apprentices, or trainees.

Above all, just as the German guilds had done, the Company was obliged to enforce high quality in all its members' products and had the power to impose harsh punishments. The Charter gave authority to the Master, Wardens and Assistants to arrange regular 'searches' to 'enter any ship, bottom, vessel, or lighters, or any other warehouse or warehouses, houses, shops, or any other place or places whatsoever, where they or any of them shall suspect any such watches, 'larums, sun-dials… and then and there to make a search for all the said clocks, watches, 'larums or sun-dials, or any other works as aforesaid peculiarly belonging to the said trade of clockmaking… and such of them as shall be faulty and deceitfully wrought, to seize on them and to break them…'

In one major respect, however, the London Company did differ from its continental counterparts. Although its byelaws allowed for the making of masterpieces, as a step in the graduation of an apprentice to a Free Clockmaker, this rule was largely ignored, and rigid masterpiece specifications never took hold. The result was a simple one. Whereas in the German states, as late as 1732, the education of a young clockmaker was geared entirely to making elaborate clockwork 'toys', to specifications drawn up a hundred and fifty years earlier, the young London makers could apply themselves to learning new technology, whenever it became available.

CIVIL WAR

Below:
A fine Swiss watch, signed 'A. Senebier A Geneve' c.1630, perhaps the property of a Royalist in exile. Attached by a silk ribbon is a tiny portrait of King Charles, with the date of his execution on the reverse. The rock crystal locket is said to contain a fragment of the King's hair.
Museum No. 18

Detail

The high cost of obtaining the Charter was financed by borrowed money, and the Company's newly formed Court quickly found itself unable to service the debt. The position was further exacerbated by frequent attacks launched in the law courts on the Clockmakers' by the Blacksmiths' Company, who had developed an obsessive dislike of the new Company and those who had founded it. Worse, the outbreak of Civil War in England in 1642 brought the entire luxury goods market to an abrupt halt. The relevant pages of the Company's Court Minute Book were left blank during the war years and only about ten clockmakers appear to have continued work. The records do not reveal whether the remainder were engaged in horology or in wartime activities, such as making armaments or serving in the 'Trained Bands', the local militia regiments. The shops in the City of London and in Westminster were closed anyway by an order of 1642 and those running them sent out to build forts and ditches around the City and to attend to the defences. Many apprenticeships ran on beyond their normal end dates, apparently to make up for the periods lost in wartime activities.

Peace returned in 1647, and under the Commonwealth a steady recovery in trade of all kinds followed. The Company's own financial difficulties were eased by the heroic action of the Fleet Street watchmaker Sampson Shelton, who paid off the mounting debts in return for reimbursement as and when the Company could afford it. As far as customers were concerned, a new wealthy class emerged of those who had profited from the war, such as bankers, army officers, lawyers and creditors. Houses were rebuilt and refurnished, estates were re-established and a relative boom followed. Between 1647 and 1660 more than forty specialist clockmakers were trading in London as well as numerous watchmakers. Many of the greatest names in English horology obtained apprenticeships at this time and in every respect the new Company was able to consolidate its position.

Below:
Sir John Branston (1577–1654), Chief Justice of the King's Bench, was a typical customer of the post Civil War clock trade. He had survived both war and attempted impeachment with his status intact. His London-made clock symbolises wealth, education and temperance.

Reproduced by kind permission of the Masters of the Bench of the Honourable Society of the Middle Temple

Actual size

Above:
Small silver watch signed by Sampson Shelton, c.1640. Shelton acted as treasurer of the group that founded the Clockmakers' Company. He later became its hero, personally paying off a debt that might well have bankrupted it.
Museum No. 1480

Top and above:
A house clock signed 'Peter Closon Neere Holburn Bridge Londini fecit' c.1655. His signature, taken from a letter to the Company dated 1642, is also shown.
Museum No. 574

CIVIL WAR | 23

THE APPLICATION OF THE PENDULUM TO CLOCKWORK

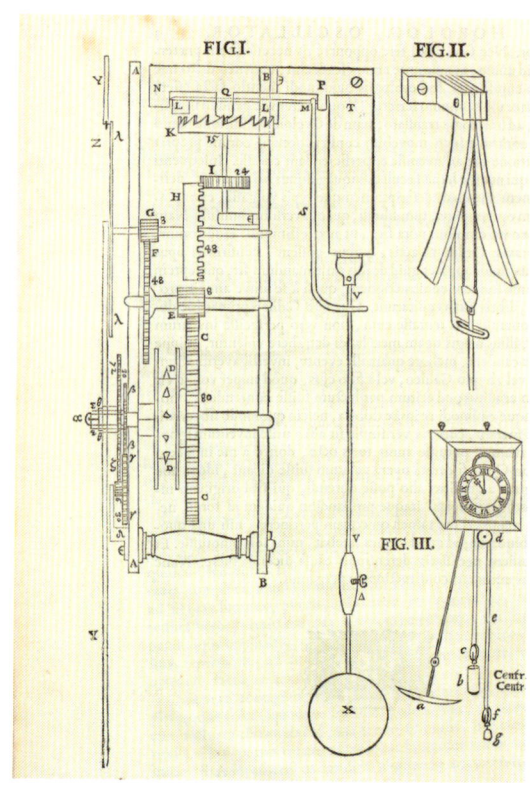

Right:
Christiaan Huygens' pendulum clock, illustrated in his work *Horologium Oscillatorium* of 1673. The particular volume from which this illustration was taken, was presented to the Company's Library by the late 18th-century watchmaker Francis Perigal.
Clockmakers' Library

The Dutch astronomer and mathematician Christiaan Huygens van Zulichem (1629–1695) claimed to have applied a pendulum to clockwork for the first time on Christmas Day 1656, thus harnessing the natural properties of a swinging body to the service of mankind.

As early as 1602, Galileo Galilei (1564–1642) had noted the same properties, that is to say, in simple terms, that a weight suspended on a given length of string or rope, takes almost exactly the same time to swing from one side to the other, no matter how great or small an arc it covers. Legend has it that the idea came to him in 1582, when he was watching the chandelier in Pisa Cathedral as it swung in the wind. He did not attempt to exploit his observation in horological terms until after he had been disabled by blindness in the late 1630s, and his own project to develop a prototype pendulum clock was cut short by his death in January 1642.

Huygens' work, on the other hand, was the greatest advance in the quest for accurate timekeeping in hundreds of years. Despite the obvious difficulties in making a pendulum that could operate at sea, the improvement in land-based timekeeping was so great that the possibility of developing a successful marine timekeeper for navigational purposes seemed only just around the corner.

Huygens commissioned first Isaac Thuret and then Salomon Coster of The Hague to manufacture his invention. It may be that at the same time Ahasuerus Fromanteel, a clockmaker of Flemish extraction working in London and one of the most celebrated members of the Clockmakers' Company, was also working towards a similar end. Whether he was or was not, it is certain that he arranged for his son John to enter Coster's service. By May 1658, with Coster's consent, John Fromanteel returned to his father's workshop with the secrets of Dutch pendulum manufacture in his possession.

That October, Ahasuerus Fromanteel placed an advertisement in the London broadsheet *Mercurius Politicus*, announcing that 'There is lately a way found out for making Clocks that go exact and keep equaller time than any now made without

24 | THE CLOCKMAKERS OF LONDON

Right:
A fine early pendulum clock in an ebony veneered case by Samuel Knibb (c.1625–1674) who became a Freeman of the Company in 1663. It was once the property of the architect Richard Norman Shaw and was presented to the Company by his son.
Museum No. 559

this Regulator (examined and proved before his Highness the Lord Protector, by such Doctors whose knowledge and learning is without exception)… You may have them at his house on the Bank-side in Mosses Alley, Southwark, and at the sign of the Maremaid in Loathbury, near Bartholomew lane end.'

At last the strands had begun to come together, for there was no better place than London for the invention to be exploited and improved. The German trade had become stifled by its own guild restrictions. The French trade was failing fast, its strength leached away by continuing religious and civil strife. The London makers had confidence, experience and a strong trade association in the Clockmakers' Company, which (whether they liked it or not) kept the quality of their work under constant supervision.

In 1660, King Charles II and his court returned from exile in Europe, offering a new and more lucrative market than any the clock and watchmakers had known since the Civil War. The king's personal interest in the arts and sciences led to the foundation of the Royal Society and of the Royal Observatory at Greenwich. The purpose of the Observatory was to chart the skies for use in navigation. Since horology was an integral part of the study of astronomy and of navigation, clockmaking was inevitably propelled once more towards the top of the scientific agenda.

Below:
The signature of Ahasuerus Fromanteel, who introduced the pendulum clock to England. Fromanteel's letter to the Company, dated 3rd of March 1656, was written from 'Mosses alley' on the Bankside in Southwark. Nearby had stood the late William Shakespeare's Globe Theatre.
MS 3942/5

THE APPLICATION OF THE PENDULUM TO CLOCKWORK | 25

THE PLAGUE

Other members of the Clockmakers' Company, such as Samuel Knibb and Edward East, quickly followed the Fromanteel family by producing both spring- and weight-driven clocks regulated with pendulums. The late Dr John Leopold showed that East, John Fromanteel and John Hilderson (who took apprentices through the Company, but oddly never became a freeman himself) became involved in attempts to adapt the pendulum to marine work. These included sea trials, Hilderson's clock being taken on a voyage to Lisbon by Captain Robert Holmes in April 1663.

But just as in 1598, 1603 and 1642, when English horological promise and enterprise had been cut short by plague and war, so the flowering of London as the clock and watchmaking centre of the world was delayed again, this time in 1665, by plague. Thomas Vincent, in his *God's Terrible Voice in the City* wrote:

'Now the citizens of London are put to a stop in the career of their trade. They begin to fear whom they converse withal and deal withal, lest they should have come out of infected places… Now death rides triumphantly on his pale horse though our streets, and breaks into every house almost where any inhabitants are to be found. Now people fall as thick as leaves in autumn, when they are shaken by a mighty wind.

Right:
A silver watch by Henry Childe who died in 1665 while in office as Master. He was a victim of the Great Plague. The outer case of the watch is covered in leather and decorated in silver piqué work.
Museum No. 1194

26 | THE CLOCKMAKERS OF LONDON

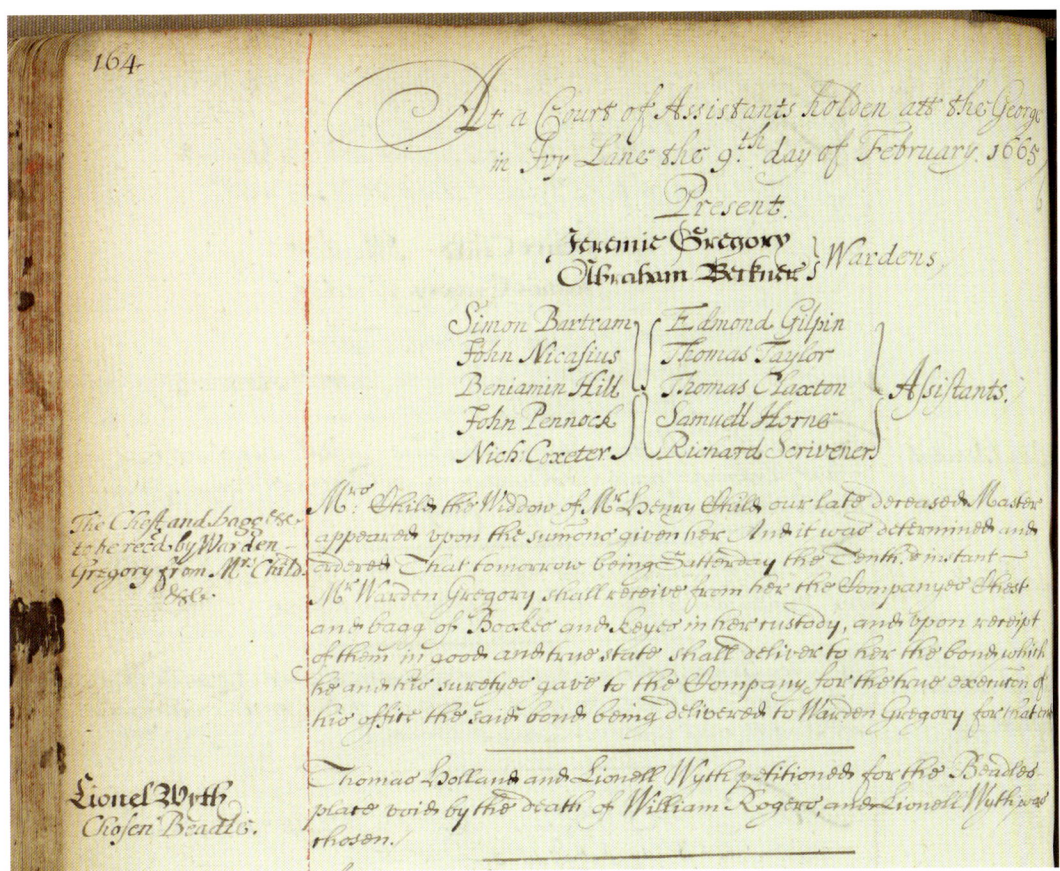

Now there is a dismal solitude in London streets, every day looks with the face of a Sabbath day… Now shops are shut in, people rare, and very few walk about, insomuch that grass begins to spring up in some places, and a deep silence in almost every place, especially within the walls, no prancing horses, no rattling coaches, no calling in customers, no offering wares, no London cries sounding in the ears. If any voice be heard it is the voice of dying persons breathing forth their last, and funeral knells of them ready to be carried to their graves.'

The Clockmakers' Company's Court Minutes Book made no mention of the horror, though a number of deaths were reported between 1664 and 1666. They included that of the Master, Henry Childe, who was certainly a plague victim. The Company's Beadle, William Rogers, also died. A considerable number of freemen pleaded inability to pay their fines and quarterage. Their combined misfortune suggests serious problems within the trade, though there is no way of knowing how many other members fell victim to disease or were ruined by the total interruption of commerce.

Above:
The minute recording Mrs Childe's attendance at Court, to arrange the return of 'Companyes Chest and bagg of Bookes and keys' following the death of her husband. They were to be given the following day to the Warden, Jeremy Gregory. The minute also records the death of William Rogers, the Company's Beadle.
MS. 2710 Vol. 1

EDWARD EAST
1602–c.1695
'Chief Clockmaker to the King'

East was born in Southill, Bedfordshire. He was apprenticed in London to Richard Rogers, a Goldsmith, becoming a free Goldsmith himself in 1628. He married Ann Bull, scion of a celebrated Fleet Street clock and watchmaking family. He soon prospered. In 1630 he subscribed to the formation of the Clockmakers' Company. He was elected an Assistant of the Clockmakers' in 1632, Treasurer in 1647 and Master in 1647 and 1653. He remained close to the Goldsmiths' Company and was elected Prime Warden in 1671. His substantial Fleet Street business may well have included banking.

Above:
Miniature portrait, believed with good reason to be Edward East c.1650. Artist unknown.
Museum No. 1261

Left and right:
The exceptionally beautiful dial and movement of a striking coach watch with alarm, signed 'Edwardus East Londini', c.1650.
Museum No. 586
Photograph left by the author

THE CLOCKMAKERS OF LONDON

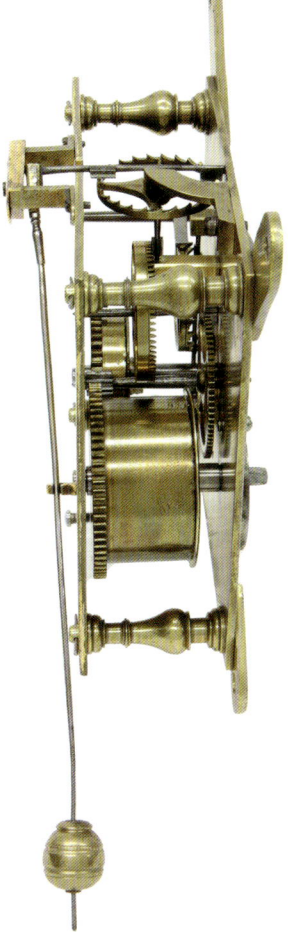

Left:
A longcase clock with an anchor escapement, signed 'Edwardus East, Londini', c.1675. The fine marquetry case, typical of London work, is perhaps ten years later.
Museum No. 544

Above:
A pioneering short-duration alarm timepiece with pendulum and verge escapement, made in East's workshop at about the time of the Great Fire. The gloriously engraved dial depicts a riot of spring flowers and retains a great deal of its original gilding. The slot at six o'clock allows the pendulum to be set in motion.
Museum No. 1267
Photographs by the author

EDWARD EAST 1602–c.1695 | 29

FIRE AND REBUILDING THE CITY

Right:
The destruction of the City of London by the Great Fire, which raged from 2nd until 5th September 1666. Rebuilding continued well into the 1680s.
Reproduced by kind permission of Museum of London

Below:
The City of London After The Great Fire by W. Hollar. The majority of houses had been timber and so were reduced to ash. Stone walls, church towers and chimney stacks were all that remained.
London Metropolitan Archives (City of London)

Worse was to follow, for on 2nd September 1666 the 'Great Fire of London' broke out and raged within (and a little beyond) the City Wall for three long days. Its progress and direction are well known. Since the Company had drawn up a list of names and addresses of its members only four years before, it is possible to calculate that of the one hundred and sixty or so listed, over half had their premises utterly destroyed. The rarity of pre-1670s pendulum clocks is largely accounted for by this fact. It cannot be a coincidence either that those makers who are now celebrated for the development of the pendulum clock, such as Samuel Knibb, Ahasuerus Fromanteel, Edward East and Henry Jones, had premises which were outside the area of the fire and so were unaffected by the blaze.

Despite the ferocity of the flames and the chaos in the streets, the fatalities were

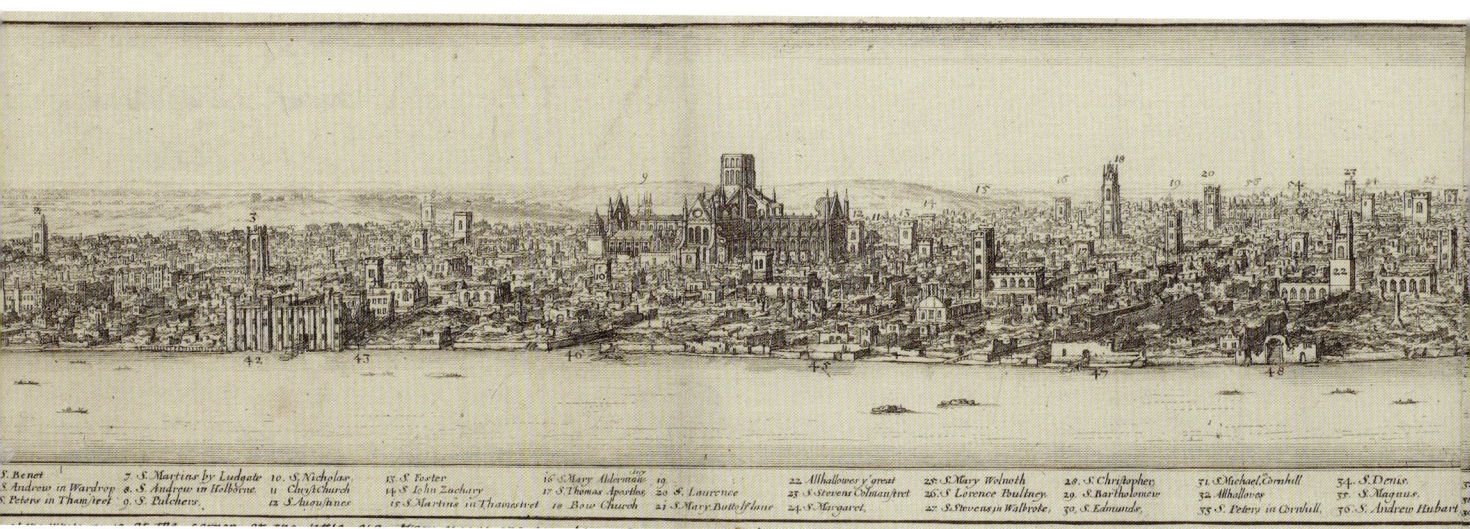

THE CLOCKMAKERS OF LONDON

very few indeed. It is extraordinary that one of these (according to an account of 1667 and reported by Bell in 1920) was an elderly member of the Company: 'Paul Lowell, a watchmaker living in Shoe Lane behind the Globe Tavern, was eighty years of age, dull of hearing and deaf to the admonition of his son and friends, who warned him of his peril. The old man declared that he would never leave his house till it fell in upon him, and kept his word, for he sank with its ruins into the cellar, where afterwards his bones, together with his keys were found.'

By a further coincidence, the only man to be executed on a charge of starting the Fire was also a watchmaker, though this time not a member of the Clockmakers' Company. He was Robert Hubert, a native of Rouen and son of a watchmaker of good standing. According to Bell, 'those who had worked with him at the bench in London and Rouen regarded him as a man of disordered mind'. Certainly, those who heard his confession were of a similar opinion and gave him every opportunity to retract. Hubert, however, went to extraordinary lengths to incriminate himself and was shortly after hanged at Tyburn.

Below:
The signature of Ralph Greatorex (1653), who had been apprenticed through the Company to the instrument maker Elias Allen. Greatorex and Sir Jonas Moore (1617–1679, Surveyor-General of the Ordnance) assessed the City following the Fire of 1666, finding that 113,200 houses had been lost. Greatorex also invented fire-fighting pumps, and was praised for this work by Samuel Pepys in his Diary (October 1660). MS2711 Vol. 1

The replanning, clearing and rebuilding of the City began at once, but was a painfully slow business, taking at least a decade to complete. During this period, London watch- and clockmaking was of necessity drastically reduced. Commercial confidence was at an extremely low ebb, not least because the hapless City felt itself at continuing risk of invasion by the Dutch, who in June 1667 sailed into Chatham Harbour. There, only thirty miles downstream of the City, they burnt or stole the greater part of the British fleet.

One member of the Clockmakers' Company did, however, benefit from adversity. He was Ralph Greatorex, who was appointed one of the two City Surveyors responsible for drawing up the first assessment of the damage. The close connection between the arts of measuring time and those of making scientific instruments and of mapping, meant that many others who were involved in the study of scientific horology were also involved in the surveying, planning and rebuilding of the City. These included Sir Jonas Moore, Robert Hooke and Sir Christopher Wren.

Probably the greatest effect of the Fire on the Clockmakers' Company as a corporate body, was subtle and unseen. Though the Company's chest containing its books, records and plate were all saved from the flames (presumably by Jeremy Gregory, the then Master), its powers were not.

City law had dictated that only those who were free of City companies could work within the City boundaries. But after the Fire it became obvious that there were insufficient builders, plumbers, carpenters and craftsmen of all sorts to deal with a disaster on such a scale. Delay in reconstruction would inevitably lead to the rapid diminution of London as a commercial centre, and the City realised that it was necessary to take drastic action. The Rebuilding Act of February 1667

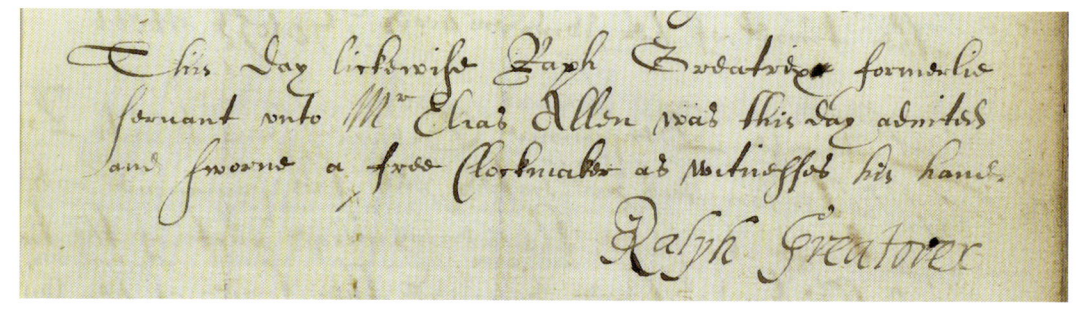

cut directly through the old traditions of protectionism, and encouraged outside craftsmen to flood in. That the Act touched companies as far removed from building work as the Clockmakers', is illustrated by the Court minute of 19th February 1673: 'Thomas Coxeter, a bricklayer, was admitted and sworn a ffree Clockmaker upon Redemption by vertue of an Order of the Maior and Court of Aldermen… he having bin imployed in the rebuilding of the Cittie of London…'

Many craftsmen made no such effort to obtain their freedom. They simply set up in the newly developed areas just west of the City boundaries, which were effectively beyond the companies' reach. Despite many attempts over the ensuing century to regain their authority, the City companies generally never succeeded in dominating their respective trades as they had in the pre-Fire years.

Left:
A survivor from the flames: an engraved 16th-century silver and silver-gilt cup given to the Clockmakers' Company in 1655 by William Petit, a watch case maker. Together with the Company's early records, it was in the care of the Master, Jeremy Gregory, when on Monday 23rd September 1666, the Great Fire engulfed his house and workshop in Cornhill. He presumably arranged for its removal through the crowds of fleeing citizens.
Museum No. 1356

FIRE AND REBUILDING THE CITY | 33

'THE GOLDEN AGE OF ENGLISH CLOCKMAKING'

Despite this adversity, the stage had been set for prosperity in the London horological trade and as the 1670s drew to a close, depression was once again followed by boom. The exploitation of many new inventions and improvements, which were largely the work of the London makers, did much to contribute to this.

The most significant of these were the 'anchor' or 'recoil' escapement, used in clocks and the 'balance spring' used in watches. The anchor escapement appears to have been devised between 1665 and 1670. It was a type of escapement which allowed a longer (usually seconds-beating) heavier pendulum to be applied to clockwork, reducing many of the instabilities in the original short, fast-beating version devised by Huygens. Its invention has been claimed for a number of clockmakers, including Joseph Knibb (a free brother of the Clockmakers' Company) and William Clement. Since its authorship was in dispute before 1700, it seems unlikely that the matter will ever be resolved. The invention is most commonly used in the familiar form of longcase clock, its seconds beating pendulum contained within its long wooden trunk. With it came a significant improvement in land-based timekeeping.

Right:
The admission of William Clement (1677), who may have invented the anchor escapement.
MS2710 Vol. 1

Right:
The minute recording the admission of Joseph Knibb to the Clockmakers' Company (January 1670–1671). Knibb was one of the finest 17th-century makers. The invention of the anchor escapement has sometimes been attributed to him.
MS2710 Vol. 1

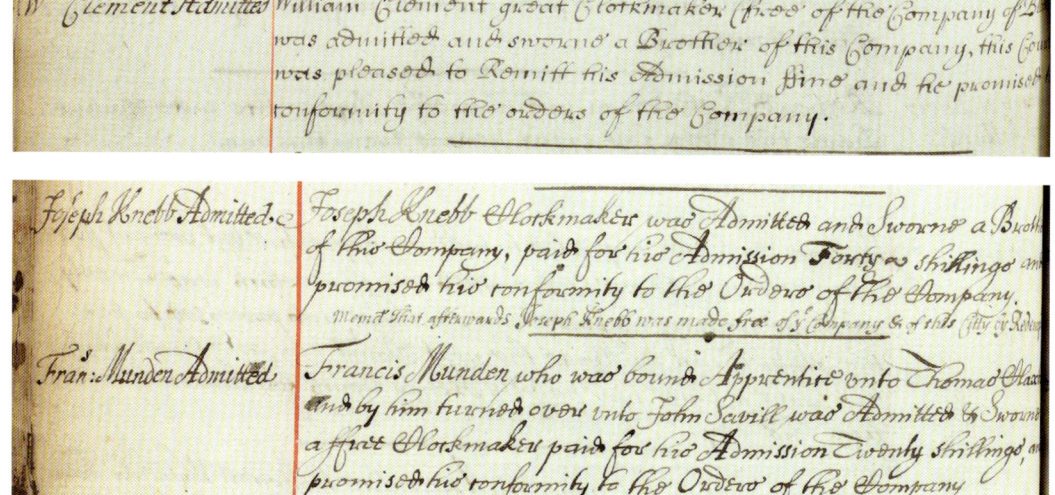

Left:
An early 18th-ccentury clock with an 'anchor' escapement by Samuel Watson. The dial displays both astronomical and astrological information. The clock is said to have belonged to Sir Isaac Newton (1642–1727).
Museum No. 564

Right:
A year-going longcase clock with an 'anchor' escapement by Daniel Quare, c.1690. The clock was originally purchased by Thomas Lord Coningsby (1656–1729) while reordering his house in Herefordshire for a visit of King William III.
Museum No. 1264

The authorship of the balance spring in watchwork is similarly disputed. It first came to public attention when Christiaan Huygens sent a cryptic message to the Royal Society in London in 1675, intimating that he had invented a method of achieving isochronism in watches, just as his pendulum had done in clocks. His right to the invention was however hotly disputed by the London scientist Dr Robert Hooke, who claimed that he had conceived the idea in 1658, had set about patenting his invention in 1663 and had lectured to the Royal Society on it in 1665. To prove his point, he immediately commissioned

'THE GOLDEN AGE OF ENGLISH CLOCKMAKING' | 35

Right:
An extraordinary survival, the menu for the Clockmakers' Midsummer Quarter Court of 4th July 1692. Amongst those who sat at the Company's table to enjoy mutton and cauliflower, beef, goose and fowl were some of the most celebrated artists of the 'Golden Age'. They included Henry Jones (then Master), William Knottesford, William Clement, Thomas Wheeler, Edward Stanton, Joseph Windmills, Thomas Tompion, Charles Gretton and William Speakman.
MS2715/2

Below:
A watch with alarm-work c.1695, signed 'Charles Gretton, London' and numbered beneath the dial '752'. The pierced silver case is stamped 'ND', probably for Nathaniel Delander. Gretton was elected Master of the Clockmakers' in 1700. Delander became an Assistant in 1689.
Museum No.68

Detail

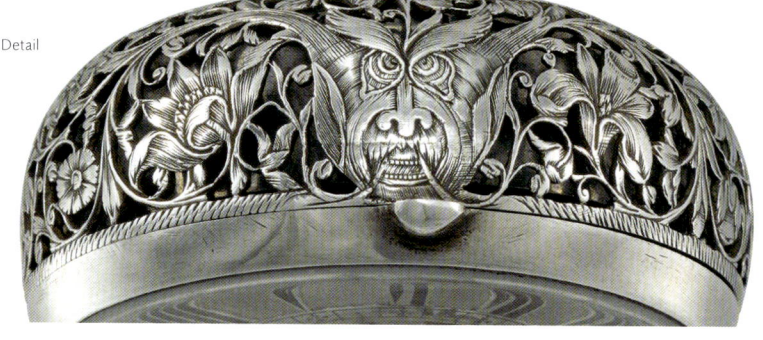

Thomas Tompion (1639–1713), who had become a free brother of the Clockmakers' Company in September 1671, to make a watch with a balance spring to his design. The watch no longer survives, so it is not possible to judge whether Hooke and Huygens had developed a similar invention independently. The discovery in 1991 of an unfinished application for a patent in Hooke's name of c.1663 suggests, however, that some form of this remarkable discovery did emanate from London. It is still in use in every mechanical watch today and played a vital role in perfecting the marine timekeeper in the following century.

These two inventions, combined with others made in London, such as repeating work and the use of jewels to reduce friction in watches, led to the production of clock and watch movements of the highest sophistication within the capital. Since the continuing influx of Huguenot refugees to the City encouraged ever-improving skills in silversmithing, goldsmithing and engraving, together with similar advances in cabinet making and design, the London horological trade, under the watchful eye of the Clockmakers' Company, enjoyed a period of prosperity which has since become known as the 'Golden Age'.

The great makers of that period were all elected to office in the Company in due course. They included Joseph Knibb (an Assistant, 1689), Henry Jones (Master in 1691), William Clement (Master in 1694), Edward Staunton (Master in 1696), Charles Gretton (Master in 1700), Joseph Windmills (Master in 1702), Thomas Tompion (Master in 1703) and Daniel Quare (Master in 1708). Their work furnished the houses and the pockets of the wealthy throughout the British Isles and Europe and continues to form the basis of many public and private collections across the world today.

By the end of the century the Company had inevitably built up respectable funds.

Left:
A fine spring clock by Henry Jones of c.1675. The movement is contained in a walnut-veneered oak case.
Museum No. 561

It chose to invest these in 1697 by acquiring stock in the newly formed Bank of England, becoming the first City Company to do so. Thomas Tompion was a frequent visitor to the Bank in 1701, during his period as the Company's Renter Warden. When the Bank was nationalised in 1947 the Governor presented the Company with a silver bowl, to celebrate the fact that it had remained a stockholder without break until that year. It is engraved, 'Time with his scythe may sever links mature but Wisdom, Honour, Friendship – these endure. 1697–1947'.

'THE GOLDEN AGE OF ENGLISH CLOCKMAKING' | 37

THOMAS TOMPION
1639–1713
'The Father of English Watchmaking'

Right:
A mezzotint by John Smith of Thomas Tompion, after a portrait by Sir Godfrey Kneller. Tompion was elected Master in 1703. His customers included King Charles II, King William III and Queen Anne. He was buried in Westminster Abbey.
Bromley Catalogue No. 1114

Below:
A remarkably similar watch movement to that held in Tompion's hand. It is signed 'T Tompion London 0598'.
Museum No. 53

Thomas Tompion was born in 1639 at Northill, Bedfordshire, son of a blacksmith. He perhaps took advantage of the relaxation of City regulations following the Great Fire to find employment in London. He became a free brother of the Clockmakers' Company in 1671 and was soon 'discovered' by the polymath Robert Hooke, who introduced him to the great intellectuals of the day, to the Court and to the King. His superb craftsmanship, innate business skills and mechanical ingenuity allowed him to build a substantial business in Fleet Street. He was elected Master of the Clockmakers in 1703. He died in 1713.

Above:
A very high quality table clock signed 'Thomas Tompion Londini Fecit', of c.1695. The movement strikes the hours and can repeat the last hour and quarter. The striking can be silenced if required. It is contained in an ebony veneered oak case with gilt-metal mounts.
Museum No. 562

Right:
The earliest surviving watch movement by Thomas Tompion, c.1671. It predates the use of the balance spring.
Museum No. 39

THOMAS TOMPION 1639–1713 | 39

PERFECTING THE MARINE TIMEKEEPER

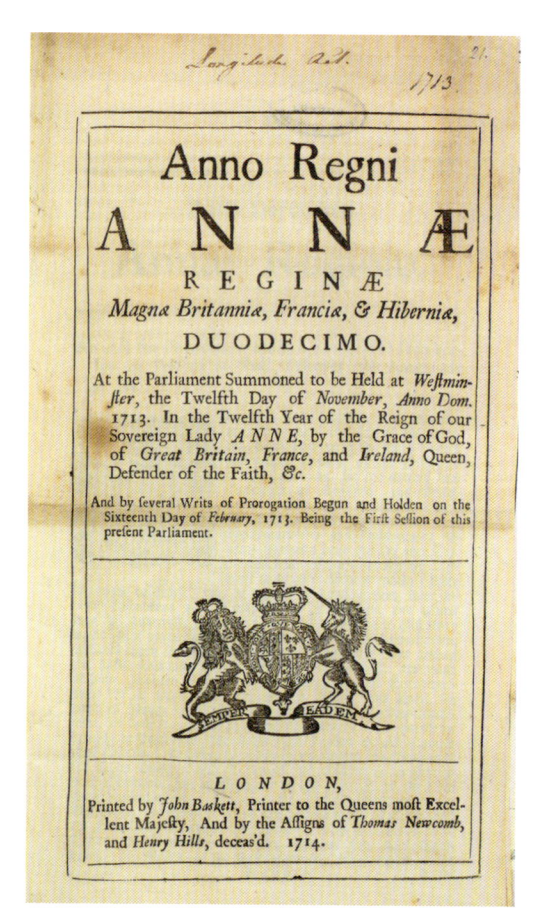

Below:
Prototype marine timekeeper by Henry Sully, 1724. Sully, who had been apprenticed through the Company to Charles Gretton, based his design on Christiaan Huygens' 'perfect marine balance' of 1693. The timekeeper failed to live up to expectations. It was presented to the Company by the clockmaker John Thwaites in 1821.
Museum no. 597

Above:
The famous Act of Parliament of 1714, which offered the 'Longitude Reward' of £20,000. This particular copy belonged to Alexander Cumming (c.1732–1814), who in 1763 was appointed an adviser to the 'Board of Longitude'. The handwritten annotation at the top of the page is his.
MS 3973/21

The need for a satisfactory method of calculating longitude at sea became so acute in the opening years of the 18th century, that by an Act of Parliament of 1714, the British Government offered rewards of up to £20,000 for a solution that would be accurate to within half a degree. This precipitated enormous interest in the subject by clockmakers, because it had long been realised that the difference (at the same instant) between the time at a ship's home port, and local time found by observation at sea, would equate to degrees of longitude east or west of that home port. A highly accurate timekeeper, capable of carrying 'home' time through storm and tempest, heat and cold, without variation, would of course be the ideal solution. Many people thought with good reason that such an accurate timekeeper could never be made, given the difficulties clockmakers had always faced with friction, poor oils, corrosion, and the expansion and contraction of metals. They offered a multitude of alternative solutions to the problem, some of which were reasonable and some of which were absurd.

Freemen of the Clockmakers' Company

Left:
'Honest' George Graham (c.1675–1751): Quaker, inventor, successor to Tompion, Fellow of the Royal Society and Europe's leading instrument maker, who earned his soubriquet largely by his generosity to John Harrison (1693–1776).
Science Museum No. 1868.249

naturally played a leading part in the race to achieve the required horological perfection. In particular George Graham F.R.S. (c.1675–1751), successor to the by then celebrated Thomas Tompion, devised considerable improvements in land-based timekeeping, with his developments of Tompion's 'dead-beat' escapement for clocks and especially with his own mercury-filled pendulum. This was designed to compensate for the effects that temperature change had on standard pendulums. Another freeman, Henry Sully, working largely on the Continent, produced a remarkable (though unsuccessful) marine timekeeper that he sent to London in 1724, its frictional-rest escapement based on Christiaan Huygens' 'Perfect Marine Balance' of 1693. Other active members included Graham's apprentice Thomas Mudge (c.1715–1794), whose work secured a £3,000 award from the Board of Longitude, the government's agents and adjudicators under the 1714 Act. The lever escapement that Mudge invented in the mid-18th century is still used in the majority of mechanical wristwatches to this day and was indeed employed in the only watches to be used on the moon.

Left:
Thomas Mudge (c.1715–1794), an oil painting by or after Nathaniel Dance. Mudge became a Freeman of the Clockmakers' Company in 1738 and a Liveryman in 1766. The lever escapement which he invented is used in most mechanical watches to this day.
Bromley Catalogue No. 1083

    The man who would eventually create a successful marine timekeeper and so receive the Government reward was not however a member of the Company, nor even a trained clockmaker, but John Harrison (1693–1776), the son of a Yorkshire carpenter. Harrison's extraordinary ability to think laterally and work around apparently insoluble problems, led to spectacular advances in temperature compensation and reduction in friction. In this he was much encouraged by George Graham, to whom he confided his ideas some time around 1728. His work took a lifetime to bring to perfection. It brought him to live in London and inevitably he became involved with other freemen of the

Below:
An important pocket chronometer by Larcum Kendall. Kendall is principally famous for the three versions he made of Harrison's prizewinning watch 'H4', one of which sailed with Captain Bligh on the ill-fated *Bounty*. This smaller watch (made c.1787) with its distinctive spiral bi-metallic compensation curb was bought from Kendalls' estate by the Royal Clockmaker Benjamin Vulliamy for £30. His son, B.L. Vulliamy, presented it to the Company in 1849.
Museum No. 422

42 | THE CLOCKMAKERS OF LONDON

Clockmakers' Company, including John Jefferys (1701–1754) and Jefferys' apprentice, Larcum Kendall (1719–1790). Harrison was in his 79th year before he received the final payment from the Board of Longitude. He was a giant among scientific thinkers of his day. The oil-free bearing, the caged-roller bearing and the temperature-sensitive bimetal strip (adapted today for use as a thermostat in computers, cars, fridges and in many other applications), are perhaps the best known of many by-products of his horological studies.

By the time Harrison had produced his fifth and last marine timekeeper in 1770, two London rivals, John Arnold (1736–1799), who later became a freeman of the Company and Thomas Earnshaw (1749–1829), who never actually became a freeman and yet paid quarterage to the Company, were soon to perfect simpler and cheaper methods of achieving a similar result. Earnshaw's escapement of 1781 was indeed so successful that it continued in use (almost unaltered) on all ships until the 1970's.

Above:
Harrison's 5th and last timekeeper, lying in the box in which it was to be taken to sea. At sea it was to be sandwiched between the damask-covered cushions, the upper one having a central hole to allow the dial to be read.
Museum No. 598

Below:
An early example of a pocket watch by John Arnold (1736–1799) containing his pivoted detent escapement. Signed on its movement 'John Arnold Invt. et Fecit No.28', it was completed c.1777.
Museum No. 419

Below:
An important watch containing Earnshaw's spring detent escapement and signed on the movement 'Thos. Earnshaw Invt. et Fecit No. 1514'. This was one of the timekeepers carried by Captain George Vancouver on the voyage (1791–1795) that included a circumnavigation of the North American island which still bears his name.
Museum No. 427

PERFECTING THE MARINE TIMEKEEPER | 43

JOHN HARRISON
1693–1776

'The Clockmaker who changed the world'

Right:
A sensitive portrait of Harrison in enamel paste by James Tassie. This copy was Tassie's own.
Bromley Catalogue No. 1082
Photograph by the author

The Clockmakers' Museum is fortunate to own a collection of Harrison artefacts second only to that at The Royal Observatory Greenwich, together with one of the most important groupings of Harrison manuscripts in existence. On display in the museum is the earliest known wooden longcase movement by Harrison, a second wooden longcase movement, one of two regulators (signed James Harrison) made for his personal use, and the marine watch signed 'John Harrison & Son London 1770 No.2', commonly known as 'H5' (see overleaf). This unique collection is augmented by Harrison's personal pocket watch and a half-scale reproduction of his first marine timekeeper, 'H1'.

Right:
Harrison's personal pocket watch, made by John Jefferys and completed in 1753. It contains all but one of the refinements included in Harrison's prizewinning marine timekeeper 'H4' and represents the turning point in his career. It was severely damaged by enemy action in World War II hence the blackened dial and scorched movement.
Museum No. 187
Generously loaned to the Company by The Corporation of the Hull Trinity House.

44 | THE CLOCKMAKERS OF LONDON

Left:
Harrison's personal regulator, acquired by the Company in 1877. It contains three of his most important inventions: oil-free bearings, a friction-free ('grasshopper') escapement and a temperature-compensated pendulum. This was one of the two regulators Harrison kept in his cottage at Barrow and was regulated by observing the apparent movement of the stars in relation to his window frame and his neighbour's chimney.
Museum No. 553

Above:
Harrison's revolutionary proposals for a marine timekeeper, which he wrote following a meeting in London with George Graham (then Renter Warden of the Clockmakers' Company). Graham gave him dinner, advice, encouragement and an interest-free loan to continue his work.
MS 6026/1

JOHN HARRISON 1693–1776 | 45

JOHN HARRISON'S 2ND MARINE WATCH 'H5'

'... It is a marvelously fine piece of work, and the maker must have possessed great talent and ingenuity to have conceived and carried out such a beautiful piece of mechanism, not only beautiful as a machine but one... capable of producing such excellent timekeeping results... equaling that of many of our more modern timekeepers.' (Samuel Elliot Atkins, who repaired 'H5' for the Clockmakers' Company in July 1892).

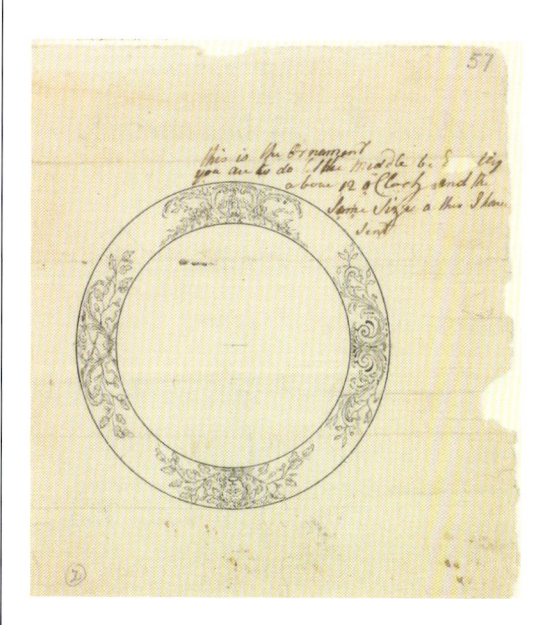

Above:
Four possible designs from John Harrison's papers for the outer ring of a watch dial. The upper one was used on his first marine watch, 'H4'.
MS 3972/3

Left and this page:
John Harrison's second marine watch, signed 'John Harrison & Son London 1770 No.2', is now known colloquially as 'H5'. Harrison made the watch to secure the last part of the famous 'Longitude Reward'. Facing obstruction, he appealed to King George III, who placed the watch on trial at his private observatory. Between May and July 1772 its average rate was less than a third of a second a day. Following a debate and in the face of Board of Longitude opposition, Parliament voted Harrison £8,750. The watch is shown here without its 'pair' case. The entire watch is shown as the cover illustration of this book.
Museum No. 598

Actual size

JOHN HARRISON'S 2ND MARINE WATCH 'H5' | 47

JOHN ARNOLD
c.1736–1799

'A skillful and ingenious workman and a clever man of business'

Left:
Portrait of John Arnold, his son John Roger and his wife Margaret, by Robert Davy c.1784. J.R. Arnold became a freeman of the Clockmakers' Company in 1796, serving as Master in 1818. He was an original benefactor of the Clockmakers' Museum.

Reproduced by kind permission of the Science Museum/Science and Society Picture Library

Below:
This marine chronometer signed 'John Arnold & Son London No.92 Invt. et Fect' is almost identical to that shown in John Arnold's hand in the Arnold Family Portrait.

Museum No. 604

48 | THE CLOCKMAKERS OF LONDON

John Arnold was the son of a Cornish clockmaker. Having spent some time on the Continent, he set up in business in London in the 1760s and there employed his exceptional talents in the scientific refinement of precision timekeepers. He took his son John Roger Arnold as an apprentice in 1783 and into partnership by 1784. He later sent him to work with the great French watchmaker Abraham-Louis Breguet (1747–1823), with whom he had struck up a remarkable friendship. Arnold became a freeman of the Clockmakers' Company in 1783 and a liveryman in 1796.

Above:
A watch in a silver-gilt case signed 'John Arnold Invt. et Fecit No. 28' and hallmarked 1776-7.
Museum No. 419

Below:
This pocket terrestrial globe by Cary, London c.1794, was John (or John Roger) Arnold's personal property.
Museum No. 1270

Above and left:
One of two late 18th-century pistols engraved with the monogram 'JRA' for John Roger Arnold. This one is flintlock with a gold pan and is signed 'I. Wilkins'. A manuscript ledger now in the Clockmakers' Library indicates that John Roger Arnold took the pistols with him for protection, when delivering chronometers to the great naval ports.
Museum No. 1275
Photograph above by the author

Detail

THE ROLE OF THE CLOCKMAKERS' COMPANY IN THE 18TH AND 19TH CENTURIES

Below:
The Clockmakers' Company produced this watch of c.1675 by Ignatius Huggerford before a Committee of the House of Commons in 1704, to defeat a patent application for jewelling watches. They falsely claimed that the purely decorative endstone in the balance cock proved prior use. They won their case.
Museum No. 45

It has been seen that the Clockmakers' Company's right to absolute control over its trade had begun to diminish as a consequence of the Great Fire. Nevertheless, the Company remained a powerful influence at the end of the 17th century and the beginning of the 18th. It saw its raison d'etre as being to root out bad workmanship and to protest against excessive or dishonest imports of foreign work (including foreign forgeries bearing spurious London signatures). It fought against what it regarded as unfair regulation in the use of precious metals, which ensured that London-made watches were expensive in comparison with imported goods. It opposed patent applications that it felt might prejudice the good of its members as a whole.

A celebrated, even notorious, example of its activity was its campaign in 1704 against an attempt by Nicholas Facio, Peter Debaufre and Jacob Debaufre to renew their

50 | THE CLOCKMAKERS OF LONDON

Left:
A copy of the Clockmakers' petition to the Crown against a patent application by Charles Clay of Flockton, Yorkshire (c.1717). Clay had claimed that a repeating mechanism which could be attached to pocket watches was his invention. The Company strenuously denied his authorship.
MS 3940/32

patent for the jewelling of watch movements 'not for ornament only, but as internall and useful part of the work or engine itself'. As evidence that jewels had been used in watch movements even before the first patent was granted, the Company produced a watch of c.1675 by Ignatius Huggerford, one of its members. This apparently contained a large ruby. The Company won its case and locked its watch away in its chest. It was not examined again until the early years of the 19th century, when it was found that the ruby was entirely decorative (perhaps even glass) and could never had been either an 'internall' or 'useful' part of the movement.

Company activities in many ways seem less high profile in the mid-18th century. Certainly, this was a period during which many makers working in London ceased to trouble to join its ranks. The Court believed that the problem lay in the fact that the Company lacked a 'Livery' and was therefore regarded as 'second-rate' by Londoners. This meant in simple terms that unlike most other companies, the Clockmakers' had never acquired the right to promote its senior members to the rank of 'liverymen'. Liverymen were those who were enfranchised to vote in the elections for the

Below:
A perfect example of the 18th-century London watchmaker's art: a gold quarter-repeating watch with cylinder escapement signed 'Geo. Graham London 883'. Hallmarked 1745.
Museum No. 317

THE ROLE OF THE CLOCKMAKERS' COMPANY IN THE 18TH AND 19TH CENTURIES | 51

Left:
Portrait of Benjamin Vulliamy (1747–1811), artist unknown. Vulliamy was appointed Royal Clockmaker in 1773 and was on close terms with King George III. His appointment as an Honorary Freeman of the Clockmakers' was something of a coup for the Company.
Bromley Catalogue No. 1086

overall government of the City and who were entitled on occasion to join in City ceremonial, dressed in the 'livery' or colours of their Company. Much time and trouble was expended from 1748 onwards in trying to persuade the City's Court of Aldermen to extend the dignity and voting rights of a livery to the Clockmakers'. Success was achieved on 1st July 1766. Members of the Court were immediately admitted to the livery, followed within days by such celebrated freemen as Thomas Mudge, John Shelton and William Dutton. To increase its influence over the trade even further, in 1781 the Court agreed to invite a number of very influential clockmakers who had not yet joined the Company to become honorary freemen. These included Benjamin Vulliamy, James McCabe, William Frodsham, Josiah Emery, Alexander Cumming, John Leroux, John Grant and Francis Perigal.

Left and below:
Sublime 18th-century workmanship: a gold pair-cased watch signed 'Thomas Mudge. London 290'.
Museum No. 725

52 | THE CLOCKMAKERS OF LONDON

Records show that as a result of this intake of significant new members, the Company again became highly active at the close of the 18th century and in the early and middle years of the next. Areas of concern were once more foreign imports, the standard of gold in watchcases, and forgeries. In these matters the Company spoke out vociferously on behalf of the trade, though often with minimal results. Even as late as 1851, the Court declined to take an interest in the Great Exhibition, on the grounds that it would be 'prejudicial to the British Manufacturers [and] that foreigners would be able to hand out trade cards in all directions, which they had done at the Exposition at Paris'. Perhaps the Company's best-known campaign was that of 1797, when William Pitt proposed and then introduced a tax on the owning and using of clocks and watches. The effect on the London trade, which (according to the Company) numbered some 20,000 workmen, was instant and disastrous. The Act was duly repealed in 1798. Perhaps the Company should bear part of the responsibility for the fact that in the following year Pitt sought an alternative source of income, and found it through the introduction of Income Tax.

Below left:
In 'as new' condition, the inner case of a triple cased watch by Paul du Pin, London. The high relief repoussé work is signed 'Moser'. The case is hallmarked 1740.
Museum No. 157

Below centre:
Early 18th-century casemaking at its most beautiful – the outer case of a watch signed 'Willm Webster, Exchange Alley, London 1033'.
Museum No. 74

Below:
A 'dumb-repeating' watch by Conyers Dunlop, Master of the Clockmakers' Company in 1758, for Queen Charlotte. The superb case is the work of George Michael Moser, renowned goldsmith, chaser, engraver and enameller. Moser was Drawing Master to King George III, friend of Dr Johnson, Oliver Goldsmith and Sir Joshua Reynolds, and a founder of the Royal Academy.
Museum No. 319

THE ROLE OF THE CLOCKMAKERS' COMPANY IN THE 18TH AND 19TH CENTURIES | 53

19TH-CENTURY CHRONOMETERS

Vast numbers of superb quality marine chronometers, pocket chronometers, deck watches and other precision timekeepers were manufactured by members of the Clockmakers' Company and others from the beginning of the 19th century onwards. It was through their use on land and sea that the British nation was able to explore, map, trade and ultimately to build an empire.

Until the advent of electronic

Above:
William Frodsham (1778–1850), partner in Parkinson & Frodsham, chronometer makers. Master of the Clockmakers' Company, 1836 and 1837.
Bromley Catalogue No. 1101

Right:
This watch by Parkinson and Frodsham was used for navigation in Captain William Parry's attempt to reach the North Pole in 1827. Parry reached 82° 45' North before being driven back, a record which stood for almost 50 years.
Museum No. 447

Right:
This famous watch, made c.1794 by John Arnold, was used 1840–1940 by John Henry Belville, his wife Maria and ultimately his daughter Ruth (seen here), to carry Greenwich Mean Time to London and to sell it to chronometer makers and others in need of it.
Museum No. 429
Photograph of Ruth Belville by kind permission of Graham Dolan

communication, it was difficult for chronometer makers who worked at any distance from Greenwich Observatory to establish the exact time. This was required for the purposes of regulation. However, in around 1840, John Belville, superintendent of the Greenwich Chronometer Department, conceived the idea of carrying the time to them, for a fee. He thus created a family business, selling time.

Left:
Marine chronometer 5338 by Victor Kullberg of Islington, first issued to *HMS Rambler* in 1894 and *HMS Pegasus* in 1905. In 1910 it was taken on board *Terra Nova* as one of eight precision timekeepers accompanying Captain Robert Falcon Scott on his ill-fated expedition to the Antarctic. It later served on *HM Transport Polgarth*, *HMS Calliope* and *HMS Enchantress*.
On loan from the National Maritime Museum

Centre page:
Herbert Ponting's famous photograph was taken at Cape Evans in January 1911. *Terra Nova*, with Kullberg 5338 on board, lies anchored in the distance.
Scott Polar Research Institute, University of Cambridge, with permission

Below:
Precision for domestic purposes: a superb mantel chronometer by Victor Kullberg, one of the most celebrated recipients of the Clockmakers' late 19th-century chronometer prizes.
Museum No. 702

19TH-CENTURY CHRONOMETERS | 55

MEMBERSHIP OF THE COMPANY

Above:
A print from a long-lost copper plate of the Company's arms, attributed to the engraver Simon Gribelin (1661–1733).
Clockmakers' Library

Above:
Portrait of Elias Allen (died 1653), engraved by Hollar. Allen became a freeman of the Clockmakers' Company in 1633 and was elected Master in 1636. He was the leading scientific instrument maker of his day, but made neither clocks nor watches.
Science Museum/Science and Society Picture Library

It is natural to assume that all freemen of the Company were directly involved in the manufacture of clocks and watches. In the mid-17th century this was so, although, as has already been seen, the power of the Company extended over a wide range of related specialist trades such as case making, sundial making, mathematical instrument making, and engraving. Richard Morgan, who had petitioned the King in 1631 for the Clockmakers' Charter, was by profession a spring maker. Elias Allen, the Company's fourth Master, was a sundial and instrument maker of great distinction.

Sons of freemen (provided they were born after their fathers had become free) were entitled to become freemen themselves, by right of 'patrimony'. This meant that if they wanted to work in the City, but not necessarily follow their fathers' trade, they might claim the right to work through patrimony, but actually follow a different profession. A surviving annotated list of Company members dated 1838 shows that by that time only 47 of the 135 senior members (or liverymen) had any connection with horology at all. John Haydon, whose fine portrait forms part of the Company's collection, is one such example. He obtained his freedom as a Clockmaker in 1771 but his prosperity came through his work as a haberdasher. Amongst other freemen of the Clockmakers' who have achieved lasting celebrity, but not by making clocks or watches, are the engravers Simon Gribelin (1661–1733, who became a free brother in 1686), Matthias Darly (freed in 1759) and John Basire (freed in 1756).

Simon Gribelin was a native of Blois and came to London around 1680. Between 1682 and 1705 he published four major works of engraved design of the highest quality, following them with engravings of the Raphael Cartoons and Reuben's Banqueting House ceiling. He is particularly celebrated for his exquisite work on silver and gold, though stylistic similarities suggest that he may have worked on clock back-plates for Thomas Tompion and Daniel Quare.

Darly was apprenticed in 1735, though not freed until some time after he had set himself up as a drawing master, print seller, designer, wallpaper manufacturer, caricaturist and engraver. He published four major works between 1754 and 1768, mostly of furniture design. He engraved many humorous plates and caricatures, including the 'Macaroni' series, but is best remembered as the engraver who produced the majority of the plates for Thomas Chippendale's *Director* of 1754 and 1762. Indeed, three of the four plates of clock design were engraved by him.

Above:
The Clockmakers' Museum does not own a watch by a female member of the Company. This watch of c.1685 is, however, by Anna Adamson, daughter of a Lancashire recusant, who after working as a watchmaker in London, left England to enter a convent in Paris. Her sister is known to have been a specialist maker of piqué watch cases.
Museum No. 57

Above and left:
A watch in a Staffordshire enamel case c.1790, signed 'Ann Phillips, London' and numbered 1778.
Museum No. 242

An indication of Darly's early importance in the stylistic development of the 18th century is given by Simon Jervis, who describes his premises (opposite Old Slaughter's Coffee House in St Martin's Lane) as 'a seed bed of English Rococo'. Jervis further speculates that it could have been Darly who first taught Chippendale to design.

John Basire came from a family of engravers, his father being Isaac Basire, copper-plate engraver, map engraver and printer of Clerkenwell, and his brother, James Basire, being engraver to the Society of Antiquaries. John was apprenticed through the Clockmakers' Company to James Gibbs of Whitefriars, a watch finisher, but went on to become an able and sensitive landscape engraver, working with J.M.W. Turner R.A. to produce views for the Oxford Almanac.

Although the membership has always been overwhelmingly male, women 'freemen' of the Company have long existed. Elizabeth Desbrow became the first female apprentice in 1676. Elizabeth's master was an 'attourney-at-law', freed by order of the Lord Mayor 'having taken a house and inhabiting in the new buildings of this City' following the Great Fire. The first female freeman to take a female apprentice was Sarah Hodgkinson, to whom Elizabeth Price was bound for seven years in 1701.

There is clear evidence in the Court Minutes that women who continued in business after the death of their husbands were permitted to take apprentices through the Company, just as their husbands would have been. Hannah Jones, for example, the widow of the celebrated Henry Jones, took Henry Magson apprentice in 1697, two years after the death of her husband. Widows who fell into poverty were frequently supported by the Company's charities.

The Company seems to have shown an equally enlightened attitude towards disability. The first record of this being the apprenticeship of George Harle to Samuel Glover for eight years on 14th October 1701. 'Memorandum', noted the Company's Clerk, 'that this last Boy George Harle is both Deafe and Dumb.'

Above:
A detail from 'Two designs of Table Clock-cases', plate 138 of *The Gentleman and Cabinet-Maker's Director* (London, 1754). The plate was signed 'T. Chippendale invt. et del. M. Darly sculp'.

MEMBERSHIP OF THE COMPANY | 57

THE FOUNDING OF THE CLOCKMAKERS' COMPANY LIBRARY

Right:
The London Tavern, where the Clockmakers' Company settled, held its meetings, and dined between 1800 and 1802 and again between 1851 until 1876.
London Metropolitan Archives (City of London)

Unlike many other of the City of London Livery companies, the Clockmakers' Company never succeeded in acquiring a hall of its own. This was partly because of the financial misfortunes that it suffered in its early years, and partly because it failed to capitalise on its powers to inspect and hallmark imported clocks and watches, which could have brought in an assured income for many centuries. Instead, it met mostly in premises rented from other companies. For 'Searches', when the Master and Wardens toured specific areas of the City to inspect the work of its members and to collect dues, it met in public houses. Its property, which consisted of manuscripts and plate, was kept in its 'Great Chest'. This was moved to the house of each new Master in turn (except in the celebrated case of Robert Grinkin in 1658, whose doorway proved too narrow to take it). Even if the Company had wished to build up a library, or a collection of its members' products, it would not have been able to do so, simply because it had no permanent home.

In the mid-18th century, however, a change in policy led to the Company renting suites of rooms, suitable for conducting everyday business, in substantial hotels. From 1764, for nearly thirty-six years, it had its own rooms at the Paul's Head Tavern in Cateaton Street (now Gresham Street). This was an establishment that might nowadays be likened to a conference centre.

Above and right:
From 1631 until 1817 the Company kept its written records and its plate in one of two oak chests. This one replaced the earlier version on 2nd August 1766. It has four different locks. The Master and Wardens each held a different key, thus ensuring that no individual could open the chest without the others being present.
Museum No. 741

THE FOUNDING OF THE CLOCKMAKERS' COMPANY LIBRARY | 59

When the Paul's Head closed, the Company moved briefly to the London Tavern and then to the King's Head Tavern in Poultry where it stayed for forty-nine years. In 1851 it moved again to the London Tavern, whose great dining room was described as 'the principal emporium of Turtle in the whole Metropolis'. There, many City Companies held their dinners, in surroundings that approached or perhaps exceeded Goldsmiths' Hall in grandeur.

It was in settled circumstances therefore that in November 1813, F.J. Barraud, watchmaker and son of the celebrated chronometer maker P.P. Barraud, proposed that the Company should assemble a collection of horological books. The Court took up his suggestion with alacrity and formed the Library Committee consisting of Barraud himself, Justin Theodore Vulliamy, Benjamin Lewis Vulliamy, John Jackson Jnr., Henry Clarke and Rev. Dr Robert Hamilton.

Justin Theodore Vulliamy initially took the chair, though his brother Benjamin (then only thirty-four, but who was later to become five times Master of the Clockmakers' Company, as well as Royal Clockmaker to George IV, William IV and Queen Victoria) soon became the driving force behind the project. Although B.L. Vulliamy has been vilified by some later 20th-century writers because he replaced a number of historic movements in important Royal and other clocks with his own work, the fact remains that he never missed a Library and Collection meeting from 1814 until his death in 1854, and took a scholarly interest in his subject. He frequently presented gifts to both the Library and Collection of great antiquarian horological interest, and he rightly claimed on his death-bed to have created for the Company a horological library second only in the world to his own. Other celebrated clockmakers who served on the Committee during the early years were John Jackson, Richard Webster, Edward Ellicott, William Gravell, W. J. Frodsham and Richard Ganthony.

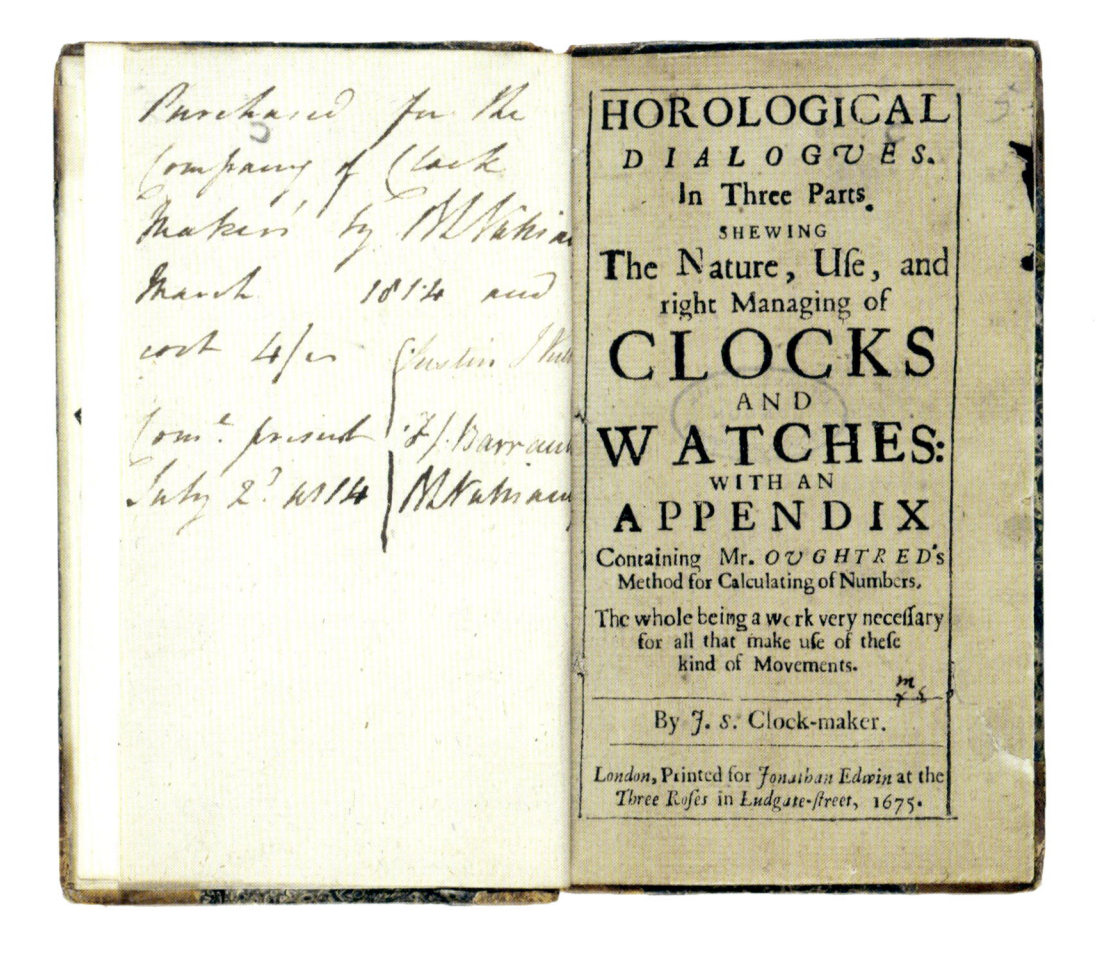

Left:
Horological Dialogues of 1675, inscribed by Benjamin L. Vulliamy and signed by 'Justin T. Vulliamy and F.J. Barraud. It had been bought for the Company by B.L. Vulliamy in March 1814 and cost 4/-.
Clockmakers' Library

THE FOUNDING OF THE CLOCKMAKERS' COLLECTION

The first meetings of the Library Committee were held at the Vulliamys' premises at No. 74 Pall Mall, the 'Asprey's' of its time. There, initially, the newly collected books were stored. Each member of the Committee presented books and soon other celebrated members of the trade, such as Matthew Dutton and Francis Perigal followed suit. Benjamin Lewis Vulliamy personally searched stalls and bookshops for rare works. One of the great strengths of the Library to this day is its collection of important books, annotated by their authors, purchasers or donors.

In December 1814, Benjamin Lewis Vulliamy informed the Committee that he had personally attended the sale of the late Alexander Cumming's property. There, on the Company's behalf, he had purchased a silver half-seconds beating watch, a pair of regulator pallets and the short duration timekeeper that Cumming had made for Capt. John Constantine Phipps' voyage towards the North Pole (1773). They were added to the Library.

Once the principle of setting up a 'library' of objects as well as a library of books had been established, watches and watch movements in particular

Above:
A watch by Debaufre, later converted to a half-seconds beating escapement. It was bought for the Clockmakers' Collection from the estate of the late Alexander Cumming in December 1814.
Museum No. 77

62 | THE CLOCKMAKERS OF LONDON

Left:
This delicate mid-17th-century enamel-cased watch signed 'Samuel Betts, Londini' was bought for the Clockmakers' Library in 1816 by the watchmaker Paul Philip Barraud. It was the fourth purchase made for the Library and cost eight guineas.
Museum No. 35
Photograph by the author

began to flood in. The Vulliamy brothers gave seventeen 'specimens of the art of watchmaking' in 1816 and the Committee noted its aim of 'procuring some of the works of the first makers, in order to form a series embracing the most distant dates possible'. It was almost certainly Benjamin L. Vulliamy who encouraged the first outsiders to make presentations. Gifts were received from Sir John Thorald of Syston Hall and The Hon. Anne Seymour Damer. Damer, a sculptress compared by Horace Walpole to Bernini, was a friend of both Nelson and Napoleon.

Left:
An engraving of the collapsible regulator built by Alexander Cumming for Capt. Phipps' voyage towards the North Pole in 1773. The spherical pendulum had belonged to George Graham. The regulator was purchased for the Collection in 1814, but was separated from the other items in 1817 because of its size. It remains to be returned to the Company.
Clockmakers' Library

THE FOUNDING OF THE CLOCKMAKERS' COLLECTION | 63

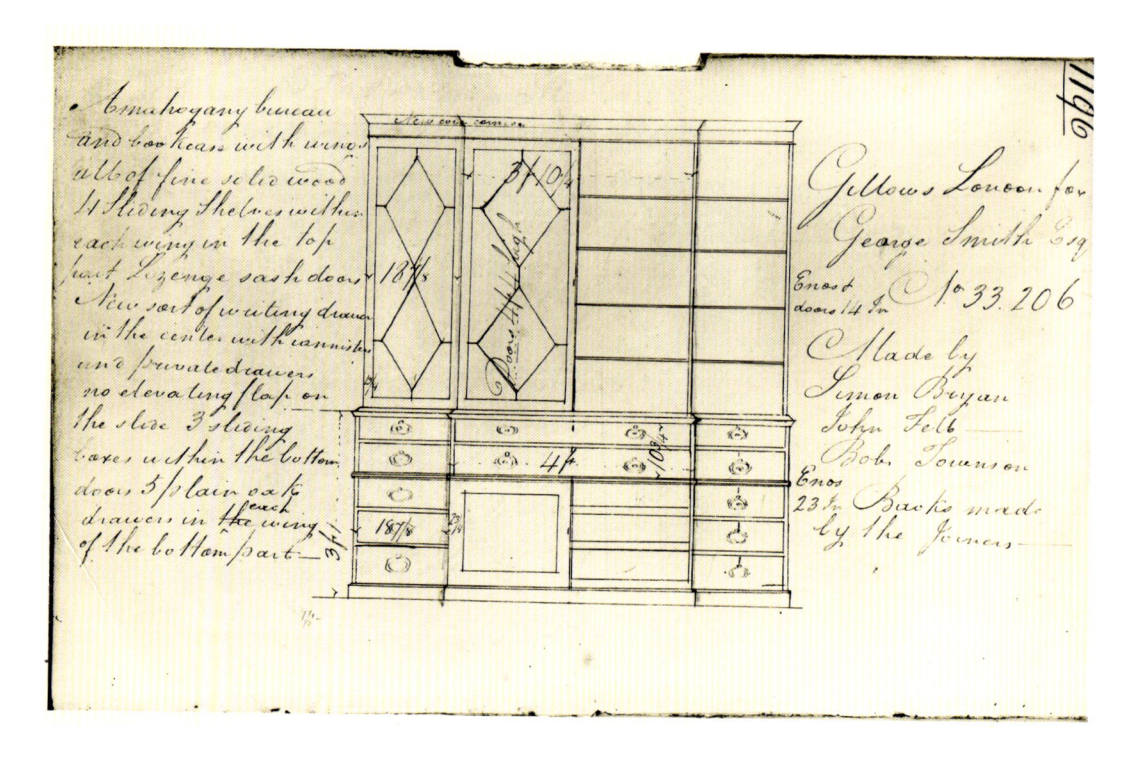

Right:
The original drawing for the bookcase bought by B.L. Vulliamy to house the Clockmakers' Library and Collection. It had been made in 1795 by Simon Bryan, John Fell and Bob Townson in Gillows' workshop and originally cost £21-10-7d.

By kind permission of Westminster City Archives.

Facing page:
The bookcase still in use today. The books on the shelves to the left and the watches on the writing slide are the very ones that B.L. Vulliamy, J.T. Vulliamy and F.J. Barraud originally placed in the bookcase on 29th August 1817.

Museum No. 1065

On 7th July 1817, twenty pounds was voted to Vulliamy to buy a suitable piece of furniture in which to store both Library and Collection. He bought with it a second-hand 'mahogany bureau and bookcase… all of fine solid wood', which had been commissioned by George Smith in September 1795 from the celebrated furniture makers Gillows of London. It was set up in the King's Head Tavern and fitted with elaborate security locks.

By 1819, five years after the Library was formed, the Committee was able to report that it had acquired 110 books, 48 watches or watch movements, 12 manuscript drawings and Cumming's short duration regulator and pallets. In addition, it held the Company's ancient records and the watch bought in 1704, to oppose Nicholas Facio's patent for jewelling. In 1821 John Thwaites, the well-known clockmaker, presented Henry Sully's Marine Timekeeper of 1724 (illustrated on page 40). It must have been a worthless object then in cash terms, but the mere fact that it was considered worthy of presentation and of acceptance, shows the depth of interest being taken in antiquarian horological matters by the founders of what remains the oldest surviving collection, specifically of clocks and watches, in the world.

THE FOUNDING OF THE CLOCKMAKERS' COLLECTION

LATER DEVELOPMENTS IN THE COLLECTION

The survival of the collection in the first half of the 19th century had depended largely on the enthusiasm of two men: Benjamin Lewis Vulliamy and George Atkins. However Vulliamy died in 1854, leaving his unique family portraits to the Company. George Atkins, chronometer maker of Cornhill and indefatigable Clerk to the Company, died in 1855. The project then lost direction and purpose. This was despite the strenuous efforts of Atkins' son and successor, Samuel Elliott Atkins, and of the watch and clockmaker John Grant, who together presented and later bequeathed to the Company some of the finest objects in the Collection.

In 1856 the Patents Office politely enquired whether it might borrow the Library and Collection complete, to be added to its own (which ultimately became part of the Science Museum, South Kensington). The Court indignantly rejected the suggestion out of hand.

Above:
One of the remarkable family portraits bequeathed to the Clockmakers' Company by B.L. Vulliamy in 1854. This painting, attributed to Jeremiah Davison, is of Vulliamy's great grandfather Benjamin Gray (1676–1764), Royal Watchmaker from 1743–1761.
Bromley Catalogue No. 1081

Below:
Benjamin Lewis Vulliamy (1780–1854), five times Master of the Clockmakers' Company, Royal Clockmaker to King George IV, King William IV and Queen Victoria, and the driving force behind the Clockmakers' Library and Collection from 1814 onwards. This wax portrait was presented to the Company by the artist who made it, Mr Lucas of Warwick Street, in July 1851. It shows Vulliamy aged 71.
Bromley Catalogue No. 1087

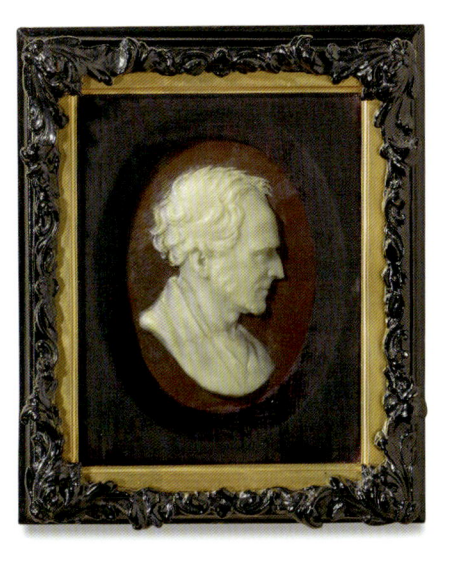

Above:
Portrait of Francois Justin Vulliamy, attributed to the Circle of Barthelemy du Pan, bequeathed by B.L. Vulliamy in 1854. Vulliamy's widow requested copies of the family portraits from the Clockmakers' Company and photographs of them were duly taken in April 1862.
Bromley Catalogue No. 1088

Above:
George Atkins (1767–1855) chronometer maker, became a freeman in 1788 and succeeded his father as Clerk to the Company in 1809. He remained Clerk until his son succeeded him in 1841. He became Master in 1845. He was an inveterate supporter of the Collection. This portrait in wax by Richard Cockle Lucas was commissioned in his honour by the Company in 1852.
Bromley Catalogue No. 1078

LATER DEVELOPMENTS IN THE COLLECTION | 67

Above:
Samuel Elliott Atkins (1807–1898), chronometer maker. He succeeded his father as Clerk in 1842, serving until 1879, during which time he never ceased to support the Collection both with his time and with generous donations. He served as Master in 1882 and 1890.
Bromley Catalogue No. 1090

Change, however, came in 1871 when John Grant, as the only surviving member of the Company's Library Committee, wrote to Samuel Atkins the Clerk bemoaning the fact that 'our Library and Museum which in former years occupied so much of your attention… has now become almost a dead letter.' He suggested that the time had arrived for proper public access and gave his opinion that both Library and Collection should be offered for display in the City's new Guildhall Library, which was then in the course of construction. The Court agreed without argument and so began not only a long and happy association with Guildhall, but also a remarkable revival in enthusiasm for the Library and Collection themselves.

The revival was led by John Grant, Samuel Atkins and William Overall (Librarian to the City), who in 1881 was to become joint author with Atkins of the definitive history of the Clockmakers' Company. They were joined in 1882 by Rev. H.L. Nelthropp, who after a period as a curate in Bristol had been posted to the British Legation in Switzerland, where he had developed a remarkable knowledge and fascination for horology. Nelthropp's greatest achievement, apart from writing his *Treatise on Watch-work, Past and Present* (London, 1873), undoubtedly lay in persuading the Court in 1891 to purchase John Harrison's fifth marine timekeeper, together with Harrison's manuscript drawings and notes. The vendor was a dealer, W. Boore of the Strand, who had purchased the collection at the Shandon Sale of 1877. The Court had turned 'H5' down in 1880 on the grounds that it was a mere copy, but now agreed to pay the substantial sum of £105 for it. Nelthropp followed this success in 1894 by presenting his entire personal collection of clocks, watches, chronometers, sundials and seals to the Company, continuing to add many other objects until his death in 1901.

In 1894 the Collection was lit for the first time by electric light. Visitor numbers increased to the extent that the Librarian noted that 'the edges of the upper cases have become much worn and rubbed, and I would suggest that… a thin iron band should be provided to protect the base of the case, as the wood is beginning to suffer from the boots of the visitors.' The inevitable consequence of this new popularity was the first loss of objects by robbery, which occurred in February 1899. The Commissioner of Police immediately provided the first burglar alarm in the form of 'a plain clothes constable to specially watch over [the] collection', an arrangement that remained in place for many years.

Samuel Atkins died in 1898, having 'attained a patriarchal age'. The numerous gifts he had given in his lifetime were augmented by a bequest of thirteen important chronometers and watches. Many other gifts and purchases were made as the century ended, by which time the Company had obtained many significant objects, from 'H5' and a longcase movement of four months' duration by Thomas Tompion, to James Harrison's ebonised longcase clock of 1728.

Above and right:
In 1849, Samuel Atkins compiled a catalogue of 'ancient watchwork' belonging to the Clockmakers' Company. He described this watch as 'a very curious small gold watch, enamell'd white in the Arabesque style, has horsehair instead of chain, hour hand only, enamell'd dial. A very beautiful specimen.'
Museum No. 32

LATER DEVELOPMENTS IN THE COLLECTION | 69

REV. H.L. NELTHROPP
1822–1901
'Ever courteous and genial'

Nelthropp began his career as a curate in Bristol, later serving as Chaplain to the British Legation in Switzerland (1851–1858). There he began to build his collection of clocks and watches. After receiving an inheritance he retired to London. In 1873 he wrote 'A Treatise on Watchwork…' and in 1881 joined the Clockmakers' Company, soon being appointed to its Court. He became much involved with its museum. In 1891 he persuaded the Court to purchase Harrison's 5th marine timekeeper, today the museum's greatest treasure. Nelthropp was elected Master in 1893 and presented his entire personal collection of 260 items to the Company in 1894.

Below:
An early 18th-century chiming table clock signed 'Johann Schmidtbaur, bamberg', bought by Nelthropp at the famous Shandon Sale in May 1877.
Museum No. 591
Photograph by the author

Above:
Portrait of Rev. H.L. Nelthropp by the artist D.A. Wehrschmidt, commissioned by the Clockmakers' Company in 1896 in gratitude for Nelthropp's generosity.
Bromley Catalogue No. 1085

Left:
Nelthropp's personal pocket watch, signed 'Geo. Blackie fecit 392 Strand London' and inscribed, 'Nelthropp Prize Winner 1873'.
Museum No. 497

70 | THE CLOCKMAKERS OF LONDON

A half-quarter repeating watch with a spring detent escapement and three-arm compensation balance, signed 'Brockbanks London No 700'. Nelthropp regarded both the movement and the case (which combine engraved three-colour gold with engine turning and translucent enamel) as perfection. He wrote 'it is hardly possible to obtain a watch on which so much time and care has been expended'. Only after 'considerable persuasion', he said, would Brockbanks' successors sell it to him.
Museum No. 440

REV. H.L. NELTHROPP 1822–1901 | 71

THE COLLECTION IN THE 20TH CENTURY

Enthusiasm for the Collection in the 20th century has been no less great. Especially well known amongst those who have taken various forms of curatorial responsibility for it are Courtenay Ilbert, whose celebrated personal collection is now held by the British Museum, F.J. Britten, whose *Old Clocks, Watches and their Makers* (first published in London in 1899) has remained almost continuously in print ever since, G.H. Baillie, author of a number of definitive reference works (including *Clock and Watchmakers of the World* London, 1929), Colonel Humphrey Quill, author of *John Harrison: The Man Who Found Longitude* (London, 1967), T. Gurney Mercer, the celebrated chronometer maker, and Cedric Jagger, son of the great sculptor Charles Sargeant Jagger, and another prolific author. Cedric Jagger's books include *Royal Clocks* (London, 1983). In 1975, Cecil Clutton and George Daniels (the celebrated watchmaker) published *Clocks and Watches in the Collection of the Worshipful Company of Clockmakers*, and in 1977 a comprehensive *Catalogue of Books and Manuscripts in The Clockmakers Library'* was compiled by John Bromley, who had been forty-four years a Librarian at Guildhall Library.

Below:
A 20th-century acquisition, a pocket chronometer by Henri Motel, its box bearing the cypher of Queen Amelie, wife of King Louis Phillippe of France. Purchased from the estate of Col Humphrey Quill in 1988.
Museum No. 1135

Actual size

72 | THE CLOCKMAKERS OF LONDON

Above:
A very fine mid-17th-century continental watch by David Margotin of Paris, in a bassine enamelled case. Purchased from Colonel Quill's estate, 1988.
Museum No. 1134

Below:
The earliest surviving self-winding watch by Breguet, Paris, which is believed to have belonged to Czar Nicholas I.
Museum No. 347

Actual size

Actual size

THE COLLECTION IN THE 20TH CENTURY | 73

Above:
A delicate watch with enamelled case by L'Epine of Paris. From the Hurle-Bradley Bequest.
Museum No. 727

Right:
Part of a collection of ten tortoiseshell frames displaying decorative watch keys from the 17th, 18th and 19th centuries. Almost every material and technique imaginable was used to create these miniature masterpieces. Presented by Mrs Philip Hill.
Museum No. 681

The 'Mary Queen of Scots Skull Watch'. An article published in 1840 stated that this watch was given by the Queen to her Maid of Honour, Mary Seaton, at the time of her execution. It has recently been shown that the watch is not 16th century at all, but was probably made in the late 18th century as part of the 'Romantic Revival'. It remains one of the most popular objects amongst visitors to the Clockmakers' Collection. Presented by G.S. Sanders.

Museum No. 687

THE COLLECTION IN THE 20TH CENTURY | 75

Above:
The Collection was displayed to the public in Old Guildhall Library from 1874 until 1976, when it was moved to the new Guildhall Library in Aldermanbury (bottom right).

Above:
In 2015 the Collection was moved to its present site in the Science Museum, South Kensington.
Photograph by Nick Pope

The Collection was moved to the new Guildhall Library in 1976 as the 19th-century library building had been designated for a new purpose. There a modest display was arranged, constrained by lack of funding. But by 2001 circumstances had changed. The Clockmakers' Museum had not only received the gifts of the Hurle-Bath and Hurle-Bradley watch collections, substantially increasing it in size, it had also received a generous bequest from a former Master of the Company, the late R.G. Beloe. On the strength of the Beloe Bequest, the Company was able to instruct the present author to reorder the entire display. This he did with the assistance of many others, in particular the Chairman of the Museum and Educational Trust, Christopher Clarke, the designer Martin Shirley of Visible Edge Design Consultants, and architect Iain Langlands of Bowyer Langlands Batchelor. An appeal for further funding both from individuals and corporate bodies was met with great generosity. At last it became possible to tell the story of clock- and watchmaking in the City of London 'from the earliest times to the present day', as the founders of the collection had intended in 1816.

When the lease finally expired on the Clock Room in the new Guildhall Library in 2015, the Company was delighted to be invited to move its entire display from the City of London to become the first ever independent museum sited within the world famous Science Museum in South Kensington. A group of volunteers generously assisted in the packing and moving of the objects across London and the new larger display was formally opened by the Princess Royal on 22nd October 2015.

THE COLLECTION IN THE 20TH CENTURY | 77

GEORGE DANIELS
1926–2011

'One of the greatest watchmakers of the late 20th and early 21st centuries'

Daniels was raised in London in poverty, without parental affection. Aged five, he prised open his family's alarm clock and realised that it was a metaphor for his life – moving inexorably onwards, but without outside assistance. He determined to learn horology, despite his parents' opposition. Conscription in 1944 unleashed his innate mechanical skills, and following demobilisation, he studied, while ceaselessly repairing watches. He gained access to the work of the greatest watchmakers through meeting a collector, and when it seemed that quartz technology would overwhelm traditional watchmaking, Daniels fought back. He made a series of increasingly ingenious mechanical watches in the manner of Breguet, teaching himself to make every part. Ultimately, he devised a virtually oil-free escapement, now mass-produced by Omega.

Above:
Posthumous portrait by John Walton of Dr George Daniels CBE, DSc, FSA, FBHI, Past Master of the Clockmakers' Company, holding a pocket watch of his own making.
Museum No. 1315

Right:
An electrically driven model of the Daniels Co-Axial escapement, made by Daniels himself c.1986 to demonstrate the technology to potential Swiss manufacturers.
Museum No. 1337

Right and below:
After a period of negotiation with various Swiss manufacturers, the Daniels Co-Axial escapement was adopted by Omega and the resulting watches launched at the Basel Fair in 1999. Daniels himself was supplied with fifty of the first Omega ébauches. These (with assistance from Roger W. Smith and the London engraver Charles Scarr) he cased, dialled and presented as a limited edition wristwatch to celebrate the Millennium. These are highly sought after today.
Museum No. 1255

GEORGE DANIELS 1926–2011 | 79

THE CLOCKMAKERS' COMPANY TODAY

The Worshipful Company of Clockmakers, now nearly four hundred years old, is privileged to have an active trade still attached to its name. While it is no longer in a position to govern its 'Art or Mystery' with a rod of iron as it did in past centuries, as an institution founded by Royal Charter, it remains obliged to do everything in its power to foster the arts placed under its care.

To achieve this, the Company welcomes as its members watch- and clockmakers, casemakers, mathematical instrument makers, sundial makers and engravers, together with retailers, wholesalers, antique dealers, specialist auctioneers, watch and clock collectors – indeed anyone with a genuine enthusiasm for horology. Company dinners and events allow everyone with any involvement in the trade to meet and to exchange ideas, just as members of the Company have done for hundreds of years. Successive Masters and Wardens are still chosen from among those who are, or have been 'professed clockmakers', as the 1631 Charter demands.

The Charter, together with the Byelaws of 1632, grants the Company many privileges, but also imposes obligations. It is for this reason that the Company's Court still works where it is able to protect the public interest. Current abuses that concern it include watches advertised as being manufactured wholly in Great Britain, when they are not, and watches made by manufacturers purporting to be the successors to famous British manufacturers of the past, when their claims are spurious.

Horological education is an immensely important part of the Company's business. Through its Clockmakers' Charity (Reg. No. 275380), it owns and maintains the Clockmakers' Museum, now at London's Science Museum. The Company's Charity, through the Tappenden Bequest, may sponsor apprentices, though its close relationship with the George Daniels Educational Trust means that the majority of apprenticeships are now funded by that means. From time to time the Clockmakers' Charity sponsors competitions and exhibitions to encourage its trade, as it has done since the late 19th century. From time to time too, it will make grants to other horological museums and exhibitions. The Company may also make grants towards the conservation of important artifacts related to its trade or towards the upkeep of the graves of famous past members. It awards medals to encourage both horological innovation worldwide and horological research. Its principal awards are its Tompion Gold Medal, for outstanding achievement in horology, and its Harrison Medal, for outstanding achievement in propagating horological knowledge.

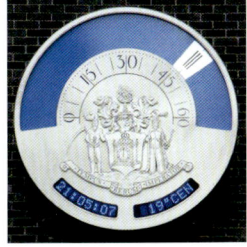

Left and above:
The Newgate Street clock, presented to the City of London by the Clockmakers' to celebrate the Millennium.

Photographs by the author

To celebrate the Millennium, the Company raised funds amongst its members and friends to present a two-metre-diameter public clock to the City of London. This clock with its remarkable 'wandering hour dial' can be seen, mounted on the brick ventilation shaft in the centre of Newgate Street, not far from the site of the 17th-century workshop of Joseph Windmills. Windmills, elected Master of the Company in 1702, was known to have made watches with similar dials. The clock was designed by Joanna Migdal, a member of the Company's Court. Smiths of Derby made the case and dial, while Keith Scobie-Youngs of the Cumbria Clock Company, another Court member, made the movement. In recent years the Company has formed an affiliation with the Antiquarian Horological Society. In recognition of the close relationship between horology, astronomy and navigation, it has also forged close links with University College London's Observatory at Mill Hill and the Royal Navy's hydrographic survey and ice patrol ship *H.M.S. Protector*. Through all these initiatives, the Company endeavours to honour its past and to look forward to the future, for it is always aware (as the motto on its 17th-century armorial bearings reminds it) that 'Tempus Rerum Imperator' ('Time rules all things').

BRITISH MAKERS OF TODAY

The Company is proud to have among its freedom and livery many clock- and watchmakers, casemakers, sundial makers and engravers, whose skills equal those of the great artists of the past. This and the following page illustrate some examples of their work: contemporary British horology at its most vibrant, ingenious and best.

This page:
George and Cornelia de Fossard, working in Wiltshire, made both the movement and the case of this complex and beautiful 'Solar Time Clock'. It contains over 750 individual pieces and took over 5000 hours to build. The movement can be adjusted for latitude and longitude, allowing it to indicate the precise time of sunrise, midday and sunset, wherever the clock is set going. Further dials show Greenwich Mean Time and local time, the moon's phase and the date.

Photographs kindly provided by the de Fossard Clock Company.

Facing page:
Prototype wristwatch movement and its constituent parts, made by Charles Frodsham & Co., London. Regulated by Frodsham's unique development of George Daniels' oilless double impulse chronometer escapement. Two independent trains deliver power to a free-sprung balance, set with four eccentric timing weights. The movement incorporates a balance brake and an up-and-down indicator.

Photographs by kind permission of Charles Frodsham & Co.

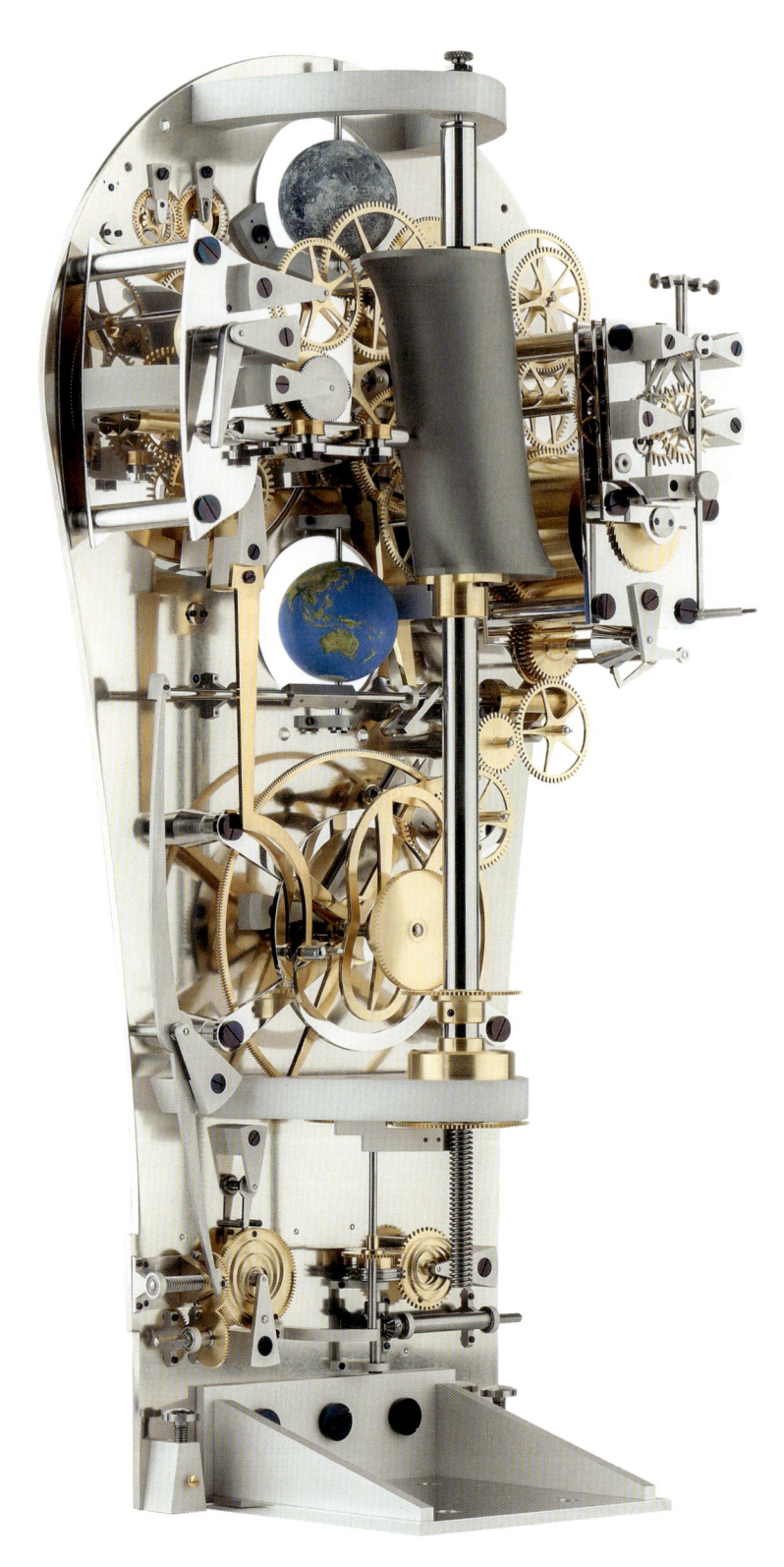

BRITISH MAKERS OF TODAY | 83

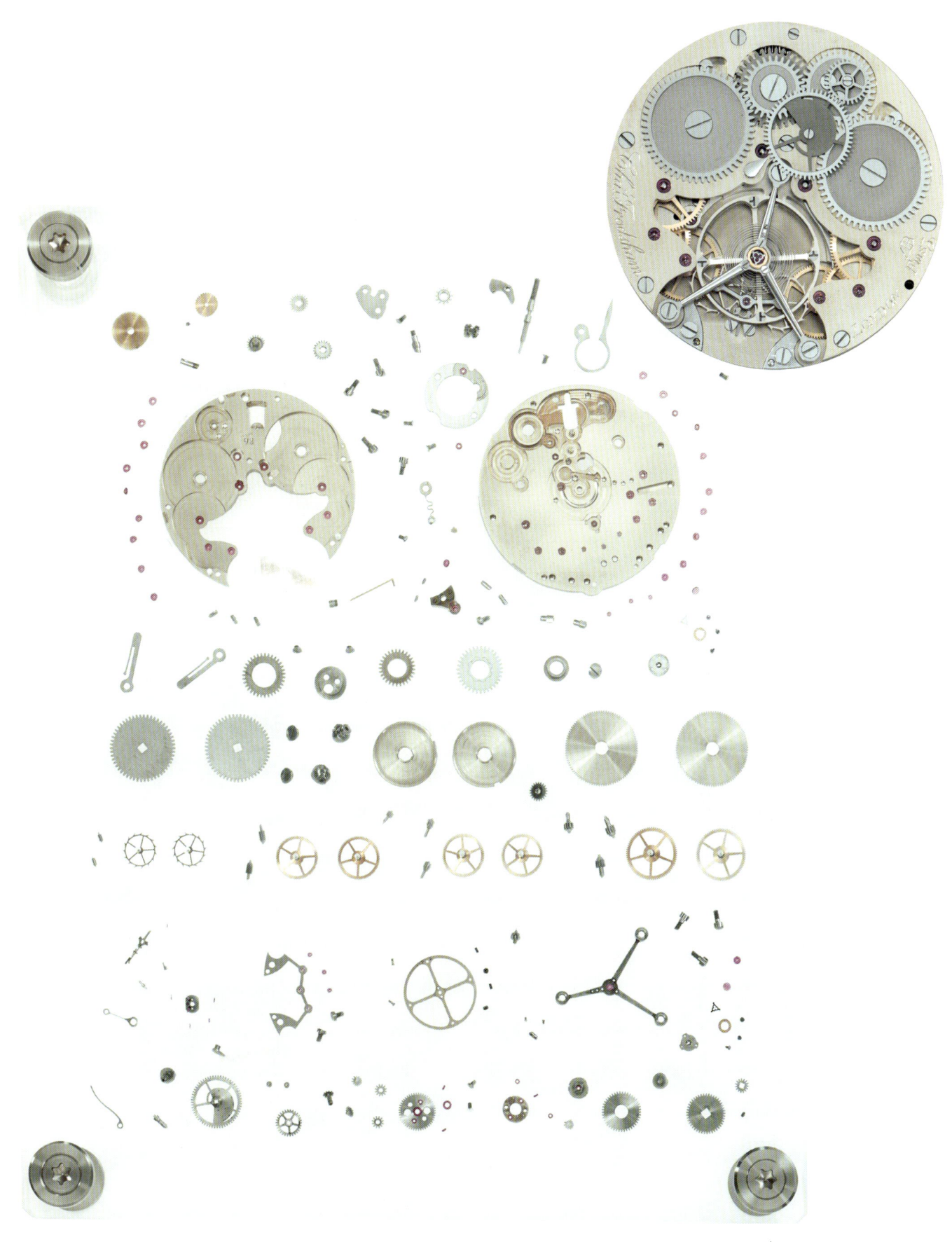

Below:
A view of the Clockmakers' Museum at the Science Museum, London.

Photograph by Nick Pope

Above:
Wristwatch in a platinum case by Roger W. Smith of the Isle of Man, made for the British Government's 'Great Britain Campaign', to celebrate British technology and innovation worldwide. The movement, with frosted gilt plates, red gold chatons, flame-blued screws and free-sprung balance, has Roger W. Smith's single-wheel version of George Daniels' Co-Axial escapement. The three dimensional engine-turned dial is cut from sterling silver with blued steel numerals and hands.

Photographs kindly provided by Roger W. Smith

Left:
The movement of a small two-train weight-driven timepiece with a Harrison grasshopper escapement, anti-friction rollers, gridiron pendulum and keyless winding by Ian Ford of Shropshire. The engine-turned silver dial shows solar and sidereal time, with up-and-down indicators showing the state of winding. The weight falls a mere eight inches in eight days.

Photograph by the author

Below:
Large and very accurate bronze polyhedral sundial, made and hand-engraved by Joanna Migdal for the Peggy Guggenheim Collection, Venice. Two dials indicate local time in summer, two in winter, while another indicates the days of Peggy Guggenheim's birth and death.

Photograph by the author

BRITISH MAKERS OF TODAY | 85

SOME EARLY BENEFACTORS OF THE CLOCKMAKERS' COMPANY

Sampson Shelton (1648)	George Graham (1747)	Samuel Fenn (1828)
Edward East (1693)	Sir Robert Darling (1770)	William Frodsham (1850)
Henry Jones (1693)	Benjamin Gibbons (1769)	Sir Jamsetjee Jejeebhoy (1855)
Charles Gretton (1701)	Devereux Bowley (1773)	Charles Rawlins (1861)
Richard Hutchings (1736)	Benjamin Sidey (1795)	William Rowlands (1864)

SOME CELEBRATED MASTERS OF THE WORSHIPFUL COMPANY OF CLOCKMAKERS

The 17th Century
David Ramsey
Sampson Shelton
Elias Allen
Richard Masterton
Edward East
Simon Hackett
Robert Grinkin
Simon Bartram
John Nicasius
Benjamin Hill
John Pennock
Nicholas Coxeter
Henry Child
Jeremy Gregory
Jeffery Bailey
Richard Ames
Benjamin Bell
Thomas Wheeler
Edward Norris
Henry Jones
William Knottesford
William Clements
Edward Stanton
John Ebsworth
Robert Williamson

The 18th Century
Charles Gretton
William Speakman
Joseph Windmills
Thomas Tompion
Daniel Quare
George Etherington
Thomas Taylor
John Shaw
Sir George Merttins
(Lord Mayor 1724–1725)
Thomas Windmills
James Markwick
George Graham
Peter Wise
Langley Bradley
Cornelius Herbert
James Drury
Nathanial Delander
William Webster
Francis Perigal
Charles Cabrier
Conyers Dunlop
Devereux Bowley
Benjamin Sidey
John Jackson
John Ward

The 19th Century
Matthew Dutton
Francis S. Perigal
Paul Philip Barraud
Isaac Rogers
John Thwaites
John Roger Arnold
Benjamin L. Vulliamy
John Jackson
Richard Ganthony
Edward Ellicot
William J. Frodsham
John Grant
William Gravell
Richard P. Ganthony
George Atkins
Charles Frodsham
John Carter
(Lord Mayor 1859–1860)
Samuel E. Atkins
Sir Joseph Savory, Bt.
(Lord Mayor 1890–1891)
The Rev. H.L. Nelthropp

Above:
A form watch in the shape of a bird, made by the aptly named partnership Soret and Jay, probably in Switzerland c.1685.
Museum No. 56

BIBLIOGRAPHY

Above:
An English 'Smiths De Luxe' wristwatch worn by Sir Edmund Hillary when he became the first (with Sherpa Tensing) to reach the summit of Mt. Everest: 11.30, May 29th 1953.
Museum No. 515

Andrewes, W.J.H. (ed.). *The Quest for Longitude* (Harvard, 1996)

Atkins, S.E. & Overall, W.H. *Some Account of the Worshipful Company of Clockmakers* (London, 1881)

Bell, Walter G. *The Great Fire of London* (London, 1920)

Betts, J. *Harrison* (National Maritime Museum, Greenwich, 1993)

Bromley, J. *The Clockmakers' Library* (London, 1977)

Camerer Cuss, T.P. (revised by T.A. Camerer Cuss) *Antique Watches* (Woodbridge, 1976)

Cardinal, C. (trans. Pages, J.) *The Watch* (Thornbury, 1985)

Clutton, C. & Daniels, G. *The Collection of the Worshipful Company of Clockmakers* (London, 1975)

Clayton, Michael. *The Collector's Dictionary of the Silver and Gold of Great Britain and North America* (Woodbridge, 1971)

Dawson, P.G., Drover, C.B. & Parkes, D.W. *Early English Clocks* (Woodbridge, 1982)

Drover, C.B. & Lloyd, H.A. *Nicholas Vallin 1565–1603* (London, 1955)

Edwardes, E.L. *The Story of the Pendulum Clock* (Altrincham, 1977)

Finley, Gerald E. *Landscapes of Memory* (University of California, 1980)

Gould, R.T. *The Marine Chronometer* (London, 1923)

Jagger, C.S. *Royal Clocks* (London, 1983)

Jagger, C.S. *The World's Great Clocks and Watches* (London, 1977)

Jagger, C.S. *The Artistry of the English Watch* (Newton Abbot, 1988)

Jervis, Simon. *The Penguin Dictionary of Design and Designers* (London, 1984)

Leopold, J.H. 'The Longitude Timekeepers of Christiaan Huygens', in Andrewes, W.J.H. *The Quest for Longitude*, p.101

Maurice, K. & Mayr, O. (eds) *The Clockwork Universe* (New York, 1980)

Quill, H. *John Harrison: The Man Who Found Longitude* (London, 1967)

Robinson, T. *The Longcase Clock* (Woodbridge, 1981)

Ronan, Colin A. *Galileo* (London, 1974)

Symonds, R.W. *Thomas Tompion, his Life and Work* (London, 1951 & 1969)

White, G. *English Lantern Clocks* (Woodbridge, 1989)

Right:
R.D. and T. Statter of Liverpool's decimal watch of 1862. The hands move anticlockwise dividing the day into ten hours. Each hour has 100 minutes, each minute 100 seconds.
Museum No. 493

BIBLIOGRAPHY | 87

Above:
Portrait of Thomas Vickery (1882–1945), clockmaker of Bridgenorth, Shropshire. Perhaps painted by the sitter himself. Signed on the reverse with the artist's fingerprints only.
Museum No. 1376

Right:
The 'Vickery Great Clock', an astronomical and musical prodigy of 1931. The astronomical observation and calculation, together with the design and manufacture of the movement, case and dial, were entirely the work of Thomas Vickery.
Museum No. 556

The Clockmakers Museum is open to the public. It is situated on the second floor of the Science Museum, Exhibition Road, South Kensington SW7 2DD. Opening times are those of the Science Museum. Details can be found on the Science Museum's website.

Correspondence concerning the history of the Clockmakers' Company and the Collection may be sent to the Consultant Keeper: keeper@clockmakers.org

Correspondence on other Company matters should be addressed to the Clerk of the Clockmakers' Company: clerk@clockmakers.org

The author is grateful to Jonathan Betts, Curatorial Adviser to the Worshipful Company of Clockmakers and to Dr James Nye for their help and advice in the preparation of this book, to Sir Harry Djanogly C.B.E., Dr James Nye and Andrew Crisford for their generous donations towards the cost of publication, to Martin Shirley, designer of this book and Rabina Stratton, who set it out.

He is grateful to Ian Blatchford, Director of the Science Museum Group, for generously providing gallery space for the Clockmakers' Museum and to all the volunteers who helped move every item from Guildhall. He is especially grateful to his wife, Joanna Migdal, sundial-maker, who gave up her work throughout the whole of 2015 to help him install the Clockmakers' Museum in its new home. It is to her that this book is dedicated.

Copyright G.S.J. White 2018

The right of G.S.J. White to be identified as the author of this work has been asserted by him in accordance with the Copyright, Designs and Patents Act 1988.

The Clockmakers of London, 2nd Edition, London 2018, published by The Worshipful Company of Clockmakers.

All rights reserved. No part of this publication may be reproduced or transmitted by any means, electronic or mechanical including photocopying or recording, or by any information storage or retrieval system, without permission in writing from the publisher and copyright holder.

Designed by Martin Shirley
Unless otherwise stated all photography by Clarissa Bruce.
Printed and bound in London by PUSH
ISBN 978-1-5272-1974-8